FUN WITh ONAKEO

Quirky Stories and Anecdotes of Snakes, Extraterrestrials and Lots of Other Interesting Creatures

FROM ARIZONA TO CALIFORNIA, TO NEVADA, FLORIDA, VIETNAM AND THE UNIVERSE

by
Richard Lapidus

GF

GOOSE FLATS
PUBLISHING
TOMBSTONE ARIZONA

FUN WITH SNAKES

Quirky Stories and Anecdotes of Snakes, Extraterrestrials and Lots of Other Interesting Creatures

FROM ARIZONA TO CALIFORNIA, TO NEVADA, FLORIDA, VIETNAM AND THE UNIVERSE

by Richard Lapidus

Goose Flats Publishing ~ Tombstone, Arizona

FUN WITH SNAKES

Quirky Stories and Anecdotes of Snakes, Extraterrestrials and Lots of Other Interesting Creatures

FROM ARIZONA TO CALIFORNIA, TO NEVADA, FLORIDA, VIETNAM AND THE UNIVERSE

by Richard Lapidus

Copyright © 2018 by Richard Lapidus
ISBN# 978-1-939345-14-1
Library of Congress Control Number: 2018937762

Published in the U.S.A.

First edition published in July of 2018

Published by
Goose Flats Publishing
P.O. Box 813
Tombstone, Arizona 85638
(520) 457-3884
www.gooseflats.com

Book layout & cover design:
Keith Davis
Goose Flats Graphics
Tombstone, Arizona
Additional images courtesy of the Keith Davis collection.

Goose Flats Publishing ~ Tombstone, Arizona

Dedication

For my amazing grandchildren:

Mary Brook, Collin Charles,

Chase Marshall and Skylar Hana

*"If there are no dogs in Heaven,
then when I die I want to go where
they went."*

Will Rogers

*"If there are no snakes in Heaven,
then when I die I want to go where
Will Rogers went."*

Richard Lapidus

Table of Contents

A Note on the Language

The small amount of profanity used in this book is part of the dialogue because it would not depict the actual stories without at least a little vulgar discourse.

In most cases I have substituted common names of reptiles and amphibians, instead of using the scientific names, the latter of which experienced snake hunters often use. It is hoped that those with and without specific knowledge of herpetology will read this work. It is clearer to name a Western Diamondback Rattlesnake than Crotalus atrox.

Animals which are named and not capitalized (i.e. kingsnake) are not specific. Where animal names are capitalized (i.e. California Kingsnake), a specific kind is identified or being discussed.

Finally, the stories in this collection are as true and accurate as possible, after the passage of up to 45 years.

HENDERSON, NEVADA
APRIL, 2018   R.L.

Introduction

I had heard there were toads on the high school football field at night. Lots of toads. Kazillions of toads.

It was reported to be Toadsville.

Only problem was, the field was locked up tight. There was no way to get in.

I was 14 years old.

I enticed a friend.* Surely we would find a way.

We rode our bikes and brought two flashlights and a big cardboard box.

It was a warm late summer night in L.A. T-shirt weather.

When we arrived at the football field, we found that it was completely dark. Then there was the matter of the ten-foot high chain link fence, which was locked at every gate.

Across the street there was the smell of someone cooking cabbage. I almost gagged. I hated that smell, and, whenever my mother was cooking it, I found some excuse to stay outside, no matter how late or how dark. We could stay outdoors after dark in those days.

There was also the neighborhood sound of a television show, one of the many westerns that dominated the airways. The volume was high, probably because the viewer was elderly and had a hearing problem. We could hear horses neighing and many television gunshots. I knew many old folks who blasted the sound on their televisions, which were still black and white at that time.

I heaved the box over the fence on the second try. Then I climbed up and over, dropping down the last few feet. My friend was chubby, and it took him a very long time.

Soon enough we were both standing on the track surrounding the football field.

We shined our lights here and there.

It didn't take long to realize that the rumors were...

Absolutely....

TRUE!!!!!!

Western Toads were hopping around everywhere on the grass.

We didn't care that the sprinklers were on!

We were big-game, bring 'em back alive, reptile and amphibian hunters! Some other time it would be reptiles. This night we were in amphibian heaven.

We filled the box until it was heavy with fat toads.

We would sell these toads and we rationalized we would better peoples' lives. We happily speculated on how many insects all of these toads could eat in just one night.

We knew that when school started in September football would be practiced and played on this field and even at our young age of 14, we knew somebody would make the decision to get rid of all of these wonderful creatures. Would the football authorities take the time to capture and relocate these creatures? We didn't think so.

So we advanced toward the fence with our heavy box. That fence looked about twice as high as it did before.

Suddenly an unmarked police car pulled up.

My first thought was to yell to my friend, "Cheese it, the cops!"

But I knew it was no use.

Someone in the neighborhood had obviously thought the school was being vandalized or looted, and had called the authorities. Maybe it was the woman cooking that cabbage. That would figure.

And now here they were.

We froze.

One of them yelled, "Freeze, motherfuckers!"

Geez, I thought. Don't they yell that when they catch murderers and bank robbers? Are we thieves? Are these toads school property?

The two men were wearing suits, but it was obvious who they were.

They were in great shape, and made it over the fence in no time. They shined their flashlights in our faces.

One of them had his hand on his gun, but he let go when he saw we were merely juvenile delinquents.

"What's in the box?" they demanded to know.

My buddy was shaking.

"Toads," I said meekly.

"What?" the second cop said.

"Toads, sir."

"Don't give us that!"

"See for yourselves," I said setting the heavy box down on the grass and opening the flaps.

They approached and shined their lights in the box.

"Jesus Christ," they said in unison.

"Look," I said and shined my light on the field, exposing hundreds of plump hopping amphibians.

The cops were stunned. They glanced at each other. They looked again inside the box of toads. They gazed back at the wet field. And they regarded us. I could see them relax a little, and I considered that a good sign.

"You boys know you're trespassing, don't you?" the first cop finally said. He was taller than the second cop, and had short blond hair.

"Yes sir." I did all the talking. My friend looked like he was about to get sick or pass out.

I asked if they thought we were breaking in to the school.

They said they were sure of it. Especially when they saw the box of loot we were carrying.

But in the end, they helped us get that box over the fence without harming any of our captives. They warned us not to do this again. They were laughing as they got into the unmarked police car, probably extremely anxious to tell everyone back at the station about their experience with the young hoodlum toadnappers.

That was my first run-in with the law while catching reptiles and amphibians. It wouldn't be the last.**

*My friend was Les Goldman. He would grow up to be Leslie Goldman, "The Enchanted Gardener." He is no longer chubby. He is a famous San Diego area poet, most well-known for planting dreams.

**See <u>Snake Hunting on the Devil's Highway</u>, "Dealing with the Authorities."

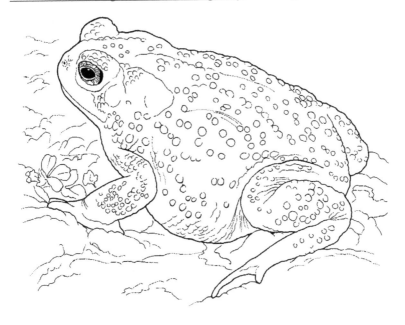

Celebrity Quote - W. C. Fields - Actor

Always carry a flagon of whiskey in case of snakebite and furthermore always carry a small snake.

Movie Quote – Anaconda – 1997

Denise Kalberg: How dare you! It was you, who brought that snake? You brought the devil!

Paul Serone: There's a devil inside everyone.

CHAPTER 1
Therapy snake makes hospital appearance

In late 1972 my friend Buz Lunsford was diagnosed with embryonal cell carcinoma, which is a medical way of saying cancer of the testicles. Early in '73 he was in the hospital trying to recover from a brutal operation, where one testicle and infected lymph nodes from all over his body had been removed. According to doctors and staff at the Sepulveda Veteran's Administration Hospital, during the surgery Buz's heart stopped twice (one said three times), and they had to bring him back. He was basically sliced horizontally in half.

A couple days after leaving the intensive care unit and being placed in a room, a cocky young intern came in for a visit. He told Buz some things about his condition that were beyond his authority to disclose. He told Buz that his cancer was an aggressive form and that the doctors were all surprised because it was most unusual for someone Buz's age to get it. His face contorted when he advised Buz that it would probably be a good idea for him to get his affairs in order, as it was generally felt that Buz's future would not surpass ninety days. If that wasn't enough, the cocky young intern said that he highly doubted that Buz would ever leave the hospital.

Buz sat up in bed as straight as possible, shook his head to clear out the medication cobwebs. He looked at the young intern and said, "Your prognosis is even worse."

"Really? Why is that?" asked the intern, suppressing a smile.

"Because I'll throw your ass out the fucking window, and we're on the sixth floor!"

The intern grabbed his clipboard and immediately reported Buz to the Chief of Staff, but the intern did not know that the Chief of Staff was a buddy who had served with Buz in Vietnam. When he came in later to have a talk with Buz, it amounted to little more than him asking, "Did you really say that?" Then they laughed and exchanged remembrances from the war.

In addition to staring death in the face and enduring a painful recovery, Buz's wife had recently left him. But he had met a pretty nurse named Sue going into surgery and, as he was being wheeled into the operating room, had asked for her phone number. Later on he told me, "Yeah, they already had me groggy and she probably figured I wouldn't remember it, if I made it out. But I did."

Several times during his recovery depression had set in. Buz got in a rebellious mood and pulled out the I.V. tubes, removed the clamps, got dressed and snuck out of the hospital. Most times he went drinking. One time, his despair was so deep, he hitchhiked close to three hundred miles north to Big Sur, a particularly beautiful area on California's central coast.

Up there he immediately cleared his lungs of the antiseptic hospital congestion and stood in the forest breathing pine-scented mountain air. Dozens of Monarch butterflies flitted about, and he watched a group of stoned-out flower children dancing, laughing, chasing butterflies. One of them, a woman with long brown wavy hair wearing a brightly-colored print dress held a small transistor radio playing "Me and Bobby McGee." Buz watched her dancing to the raspy words of Janis Joplin, smiling, almost floating, reaching for the elusive orange butterflies but never quite catching them. And not caring.

Moving along, Buz found just the right pine tree, slumped down and leaned against it with his pea coat rolled up as a pillow. He slept there that night occasionally waking to gaze up through the trees at the brightly lit night sky.

He could barely walk without pain, but over the next several days he hiked around in the mountains, traveling on foot at least twenty miles. He kept to himself, but on one occasion he came across a hippy who offered him some grass and whiskey if Buz would help carry some cookware to his campground up the trail some distance. By this time Buz was struggling to move about, but he carted the pots and pans several miles uphill, even though they were heavy cast iron.

Out of breath and panting, he told the hippy, "You know, they make these out of aluminum these days."

At the campground he was rewarded with a bag full of primo grass and a pint of Wild Turkey, both of which he readily consumed.

Ironically, while he was stoned and drinking Wild Turkey he saw real wild turkeys running through the trees near a narrow stream.

"Oh wow."

He also observed several California condors, and marveled at their huge wingspan as they sailed air currents.

The smoke and the booze eased Buz's pain, but were not potent enough to prevent him from obsessing about deteriorating in front of his kids. He had watched his mother suffer terribly from cancer, weaken and deteriorate. The hell if he would allow his four daughters see him rot like that.

He returned to the hospital when his incision became infected and the pain grew unbearable.

One day I called him and he seemed especially down. In fact I could hardly get him to talk at all. Then I thought of something. "How do you think they'd react if I brought in a snake?"

A tired, weak voice replied, "Think about it, Richard. Here's a bunch of sick vets, some just getting back from Nam where they were scared to death of snakes. Then there's the staff, most of which have probably never seen a lizard up close let alone a snake. A snake in here could actually cause a riot."

"Well," I said, "I've seen dogs brought in as therapy for patients. You know, golden retrievers, calm dogs like that. So maybe a calm snake would be good therapy for a patient like you, and if so, what's the problem? I mean, it won't fuzz up the place with fur and there's not a housebreaking issue. Could you ask someone if it would be okay? I'll just bring in a small one for a few minutes. We'll keep it in your room. Nobody needs to see it but you. It might be fun."

"Sure," he said. "I'll ask, but I don't think so."

I called him back the following morning. I couldn't tell if he sounded sick or hung-over. "Been out drinking again?" I asked, taking the odds.

"Yeah, and now they're really pissed. And they said absolutely no snakes under any circumstances."

"Okay, fine, I understand. I'll come by later. Irisse wants to come too." I hung up the phone and resolved to bring in a snake that very day.

So I started looking through the cages. I stared for a long time at the rosy boa. That's about as docile a snake as there is. Not much chance of a disturbance breaking out over a tranquil little rosy. I glanced over at a beautiful mountain kingsnake. Very pretty, but it could possibly bite, and the bright colors might scare somebody. The adult boa constrictors were extremely friendly, but they were just too big. Then I looked over at Georgia.

Georgia was a young Burmese python, about 5 feet long. She had been confiscated from a drug dealer during a bust and offered to us by the Agoura Animal Shelter. Buz and I had done a couple of lectures for their staff after they had caused a panic in their facility over the misidentification of a kingsnake they thought was a banded krait. We had brought in a few snakes, including some Southern Pacific Rattlesnakes, and had given them basic rules of identification and snake handling, and we tried to dispel as many myths as we could. Their staff was also agreeable to relocate rattlesnakes, instead of butchering them, or at least call us for assistance. It had been a win-win situation for everyone, and now we had Georgia.

Georgia was a friendly and as highly attractive a Burmese python as you could get back then, as it was years before they were being bred for albino or color morphs. She was only about five feet long, and did not yet have the bulk of a larger python. I tied her in a pillowcase, and set it down in Irisse's large animal print purse.

At the hospital, we tapped the button and the elevator lifted us smoothly to the sixth floor. When the doors opened, the nurses' station, bustling with activity, was just ahead. Irisse and I exchanged glances. Later she told me that I had this devilish sort of half grin and twinkle in my eye. She was nervous but kept her cool. At that time we had been married for only five years. She tolerated, and involved herself with my snakes and snake hunting to a point, but I had never before witnessed her participate in anything so bold and daring.

When we entered Buz's room he looked awful. "You look awful," I told him.

"You don't look so good yourself," he said.

"Yeah but I look like this every day. We brought you something, might cheer you up."

I gave Irisse a look and she opened her purse and leaned it in my direction, clearly not wanting to touch the pillowcase herself. I removed the bag and handed it to Buz, still tied on top.

"You brought it anyway. Thanks."

Immediately his mood perked up as he began to work on the knot. "Why'd you make it so tight?"

"Well, for one thing you taught me to do that. Remember that loose Mojave in the car?* And for another thing, it was in Irisse's purse. I wouldn't want to be around if it poked its head through a loose knot and came out visiting."

Buz slowly eased himself out of bed. As he unknotted the bag he said, "I see your point." Then he reached in the pillowcase and pulled out Georgia. "Oh, you brought Georgia!"

Buz's hospital room was semi-private. There was another patient in the room named Murph. His real name was Murphy, but Buz told me that everyone called him Murph.

"Is that his first name," I had asked.

"I don't know," he had replied in a tone of annoyance. "What's the difference, it's his name."

Then Buz wrapped Georgia around his neck, pulled the curtain open between the two beds and said to Murph, "Hey, look what they brought me."

Now Murph, a big skinny dude in his 50s or 60s, became animated. He sat up in bed and said, "I got a big new snake too!"

Irisse shot me a look of mild terror and I elbowed Buz. Buz laughed. "Murph over here just got a penile implant last week. They put a pump in his penis. Nurses have been coming around day and night asking if they could try it."

I started to ask, "Does...." But he cuts me off. "Yes, he lets them. He loves it, don't you Murph?"

Now Murph was smiling. He looked at me and said, "Want to see it?"

"No," I said way too loud and walked around Buz's bed to the window. Irisse joined me there immediately. That's when Sue, the pretty nurse, came in. She saw Buz with Georgia around his neck and said, "Oh wow."

Buz slowly, easily unwrapped the python and started to hand it to Sue. At this point Sue had never touched a snake. She hesitated briefly, looked at Buz and finally accepted it. Buz draped the snake around her and demonstrated how to hold it so both she and Georgia would be comfortable. Sue kept looking at it and everyone could see that she and Georgia were becoming real attached to each other.

I told Sue, "Once you start liking snakes, you can never go back."

Sue did not reply, because she was fully absorbed in holding a snake for the first time and enjoying how Georgia felt around her neck, in her hands and against her body. After a few minutes of mutual admiration, she unwound Georgia and handed her back to Buz, who had made up his mind to take a short stroll around the sixth floor. "They want me to walk several times a day," he said.

I thought *oh no, everyone is going to see Georgia and there will be a riot.*

But at the precise moment that Buz reached the door, it opened, which forced Buz and Georgia back into the room. The young intern started to come in with his stethoscope and chart. He almost bumped into Buz. That's when Georgia instantly got out of character,** hissed and lunged at the intern, her mouth open wide, just whiteness and teeth showing, lots of sharp teeth inside.

The intern was taken totally off guard, yelled out but did not really understand what just happened. He had not been bitten, but he knew that something awfully bad and scary had just occurred, and he turned on his heels and ran out.

I took Georgia from Buz, quickly stuffed her back in the pillowcase and ran it out to the car while everyone else stayed in the room laughing. I made it back to the room just seconds before the intern came cautiously back in with the Chief of Staff, the one who was a friend of Buz.

The Chief of Staff glanced around the room. Buz was back in bed now and had that pathetic depressed look about him like he had when we had first walked in. Everyone stared at the Chief of Staff. He finally said, "This man says that there's a huge snake in this room."

Buz said, "He says what?" We all looked at each other as if we didn't understand.

Georgia All Grown Up

Georgia, the Burmese Python that visited Buz in the hospital, has grown too large to fit in anyone's purse. Here Richard prepares for a hand-off to Buz.

"He says you have a snake in here and it tried to bite him."

Buz said, "There's no snake in here. If he's seeing snakes maybe he needs detox or psychiatric care."

The Chief of Staff looked at Sue. "Nurse, is there a snake in here? Sue shivered a little and said, "No sir. I'm terrified of snakes. If there was a snake, I wouldn't be standing here."

The young intern stood there turning bright red. As the Chief of Staff turned to walk out the door, Murph smiled and said, "I got a big snake. Wanna see it?"

*See *Death by Mojave*, <u>Snake Hunting on the Devil's Highway</u>

**Georgia lived for many years and grew to over eleven feet and close to a hundred pounds. She only got "out of character" one other time. See "Muffin for Dinner."

<u>Movie Quote</u>
Raiders of the Lost Ark – 1981

Indiana: Here, take this.

(hands Marion a torch)

Indiana: Wave it at anything that slithers.

Marion: The whole place is slitherin'!

Marshall Trimble
(Arizona's Official Historian)
Quirky Fact

Out in the bleak desert of western Arizona is a town called Hope. Living there doesn't dampen the spirits of its residents. They erected a sign on the way out of town that says, "If you can read this, you're beyond Hope."

You're Now Beyond Hope

Over the years there have been many different versions of this sign. This current version can be seen by motorists as they exit the town of hope

Photo courtesy of Gerald Thurman.

CHAPTER 2
The Meanest Python

The phone rang one evening in May, and Buz was on the line. His voice was louder than usual and he spoke fast.

"How would you feel about picking up a very large python?" he asked.

"I don't know," I said. "What's the deal?"

He explained that he had just received a call from Los Angeles County Animal Control. They were trying to help a woman who had suffered a serious bite from her pet snake.

My imagination immediately ran wild with images of a huge python with a tight constriction hold on a woman, its teeth firmly planted in the fleshy part of her arm.

"What kind of python? How large is very large? What constitutes a serious bite?" These were the questions I asked, but Buz had no answers. He only knew that they needed a response right away because the woman wanted the snake out of her home pronto and they had someone else to call if we didn't want it.

I told him I guessed so, and he said that's good because he had actually already agreed to take the snake regardless of species, size, disposition or condition of health.

Then I said, "You know as well as I do that people always exaggerate about the size of snakes. If it's a huge python, maybe it's a four or five foot Ball Python. But if it's a Burmese or Retic, it's probably over ten feet, and I don't have a spare cage that big."

"I forgot," he said. "The cage comes with it."

"This should be interesting," I replied.

The woman, Sally,* looked us over and then invited us into her home. We carried a heavy-duty snake stick and a huge cloth bag. Buz wore a cowboy hat and a western shirt with snaps. I wore a cowboy hat and a tee shirt with a printed python winding around the front and back.

Introductions were made and then Sally led us over to the cage and we saw the snake for the first time. It was a large, but fairly thin-bodied Reticulated Python with some nice yellow in its pattern.

Buz and I looked at each other. "Retic," we both said.

Then Sally invited us to have a seat. She wanted to tell us about the snake and why she was giving it away.

"This is not my snake," she began. "It belonged to my husband. He was a biker."

It was then that I noticed that Sally was wearing a good amount of leather and that she had several colorful tattoos. As this was in 1985, you would occasionally see some ink on chicks, but it was by no means common as it is today.

She continued her story. "But he died in a motorcycle accident a little over two months ago."

I hung my head. Buz and I each stated our condolences.

"Several years ago he selected this snake from a big litter of hatchlings. These babies were mostly biters, but he took great care to select what he thought was the most aggressive one."

Her eyes were tearing and I didn't want to interrupt, but I thought that this was like getting the reverse pick of the litter and being thrilled about it.

Sally went on. "We had the snake for a few years now. My husband took great care of it. I didn't have to do any feeding or cleaning. I did nothing but watch him enjoy the snake, and show it off to all his friends. But like I said, my husband died and I had been putting it off, but I finally figured the snake needed to eat. So I got some rabbits. I know now that it was wrong, but I set the rabbits down in a wire cage in front of the snake cage and went to work, thinking I'd do the feeding when I got home."

Buz and I shot each other a knowing glance.

"I understand now that the snake hadn't eaten in over two months, and had been staring at those rabbits all day and smelling them. So naturally when I opened the snake cage, it lunged out and grabbed my hand. I tried to pull away, but the pain was unbearable, so I let it pull my entire arm just about all the way in the cage."

At this point Sally choked a little. She got up from a dinette chair and paced around the room, finally settling in a large, overstuffed chair, and continued her story.

"Somehow I managed to reach the phone and I called 911. The firefighters arrived in about ten minutes. By then the snake was tightly coiled around my arm, and its teeth were deep in my hand. If I tried to move, he just squeezed tighter. My arm was numb and my hand was beginning to lose feeling too, which I was grateful for, believe me, but I was scared.

Python tires of menu, bites hand that feeds it

An 18-foot-long python apparently tired of the rabbits it was being fed Thursday afternoon and chose instead to sample its owner's arm.

The snake, eight to 10 inches in diameter, left several cuts on the arm and hand of its victim, a 40-year-old Canyon Country woman. The woman, who was not identified, was taken to Henry Mayo Newhall Memorial Hospital in Valencia.

A python that size could have swallowed a large dog, said a spokeswoman for the Wildlife Waystation in Tujunga.

County Fire Capt. Mike Mathis said the woman was feeding her caged pet about 3 p.m. when it grabbed her hand and wrist in its mouth then coiled itself around her arm. A friend called firefighters, who battled with the python before they managed to release the woman's arm, Mathis said.

The woman, who lives in the 26000 block of Gazeley Street, will require stitches, Mathis added.

Daily News Article – May 24, 1985

As in many news reports of incidents involving snakes, this one exaggerated several details. We can give them the benefit of the doubt about the size of the snake. A snake that is 16 feet long looks just as formidable as one that is 18 feet. But this snake was nowhere near the 8 to 10 inches in diameter that was reported. Could this snake have swallowed a large dog? Perhaps, if it had truly been 10 inches wide. This snake got tired of the rabbits it was being fed? Not at all. Firefighters battled with the python? If so it was not much of a battle, as the only injuries were to the bite victim, whose hand ended up severely scarred.

"When the first firefighter arrived, all he saw was coils all the way up my arm. He said, 'holy shit' and immediately radioed for backup. I told him about how this was my husband's snake, about how he had just died, and that I really didn't want anything to happen to it, but that if it came down to a decision between the snake and my arm. . . well. . . I said just kill the mother fucker."

This broke the tension and we all laughed.

"Before I knew it all kinds of people showed up. I think there was the one firefighter, a deputy sheriff, two paramedics and two ambulance attendants standing around not knowing quite what to do. They all decided that trying to kill the snake was too risky, but I have no idea if they thought it was too risky to me or to all of them. Finally they decided to take my arm and the snake out of the cage and put the snake on the floor.

"Believe it or not, that worked. The snake uncoiled and released my hand. The firefighter grabbed the snake behind the head and some of the others picked up the rest of it. Then they shoved it back in the cage. After that they treated my hand and a friend of mine came over and drove me to the hospital."

Sally got up and showed us her hand. It was a mass of quarter inch scars. I looked down at my hands, which combined had about a dozen similar scars, and showed them to Sally. Then Buz unsnapped his cuff and rolled up one of his sleeves and showed her his massive motorcycle accident scar. It was like the scene in *Jaws* on the boat where Richard Dreyfus and Robert Shaw compared their shark bite scars and Roy Scheider looked on with a smirk.

"I should have had plastic surgery," Sally said. "I still might."

"Nah," said Buz. "Just think of all the stories you can tell now. Your hand has character."

"And don't forget," I added. "You've got some snake in you now. Use it to your advantage."

Some weeks later a group of five of us stretched the python out and measured it. Sixteen feet, four inches is what we came up with.

It was a voracious eater. We kept it for several years and never had a problem with it, though we took all precautions. When it was feeding time, it was fed first, and we never allowed the sight or smell of feeder stock to permeate the room.

The firefighter's idea to take the snake out of the cage and set it on the floor was a good one. Buz and I had done that many times with biting snakes and lizards, and it almost always worked.

*Not her real name

The Meanest Python – Hold That Head

Slightly shy of the 18 feet length reported in the newspapers, this Reticulated Python was selected by its owner because it was especially aggressive. In a feeding accident, the snake seized a 40-year-old woman's hand, and wrapped itself around her arm. Firefighters and paramedics were called, and they eventually freed the woman and returned the snake to its enclosure. The woman was the widow of the man who wanted the meanest python. Her hand was badly scarred from the incident. Here the snake is being stretched out for measurement. Left to right: Buz, Richard, Max Peterson, Unidentified, Billy Davis

Petroglyph in the American West

CHAPTER 3
Bisbee: Copper Pits and Pit Vipers

When you're collecting snakes on Cochise County, Arizona roads at night, you need to catch up on sleep during the day. When you wake up, you can search for diurnal specimens or you can spend some time visiting some of the unique areas in southeastern Arizona. Over the years Buz and I have checked out all the old west ghost towns, like Dos Cabezas, Pearce and Charleston. We've hiked all the trails through the Chiricahua Mountains. We've driven to Tombstone and explored the O.K. Corral, Birdcage Theatre, Boothill and Schieffelin Hall; and we consumed a few beverages at the Crystal Palace Saloon. However, one of the more interesting day trips we took several times was to the picturesque copper mining town of Bisbee.

When you first arrive, you'll instantly know that this place is absolutely unique, with its colorful Victorian homes, hotels, shops, galleries and shacks built right into the hillsides, many of them remarkably preserved or restored from their 1880s beginnings. But the first thing you will see when entering Bisbee from the east along Highway 80 is the famous Lavender Pit open copper mine, where thousands of tons of copper ore and waste rock were removed daily prior to 1975.

A literal mountain of copper ore was removed, as the big hole in the ground was once known as Sacramento Hill and towered high above the town.

When visiting Bisbee drive slowly around the area of the Lavender Pit open copper mine, as it is a known speed trap, and be sure to pull over to see the huge hole in the ground that used to be a mountain peak.

We were driving west on Tombstone Canyon in Bisbee, Arizona. We had passed the historic district with its colorful shops, hotels, restaurants and galleries and were heading toward the Circle K at the west end of town, where Buz could procure a pack of cigarettes and a bottled diet Dr. Pepper, and I could get a 64 ounce Polar Pop Coca Cola. (I don't remember what they were called back in those days, but they are Polar Pops now.) As we approached the convenience store we noticed a small crowd of assorted bikers, hippies and tourists coincidentally standing in a circle in the Circle K parking lot. They were looking down at the ground, and two or three of the group had just lunged backwards.

"Something is afoot at the Circle K*," I said. Buz was already out the door, and I stopped the car in the middle of the lot. One of the bikers, a portly dude clad in leathers although the temp had to be 100, held a rock high overhead. I figured this stance was most likely reptile related, as we had participated in a similar event out on the famous Ajo Road a few years before.** I grabbed a snake hook and made my way over to where Buz was speaking to the biker with the rock.

"Hey," Buz said to the biker. "Where did you get that rock?"

The biker lowered the rock and took a look. I saw a coiled and rattling mid-size Western Diamondback Rattlesnake on the pavement, hot and occasionally striking, causing the circle of people to gyrate like the wave at a baseball game.

Something was Afoot at this Circle K

A small crowd of locals, bikers and tourists gathered around a rattlesnake to the right of the store near these motorcycles. Buz and I intervened, and thankfully saved and relocated the badly frightened rattlesnake.

The biker was breathing heavily and gasped for breath. "I just picked it up, why, it's just a rock."

Buz said, "Well I hate to disagree with you, but it's far from just a rock. You see that blue streak?"

Now the crowd turned from the snake to Buz to the rock.

"Why?" the biker replied with a frown. "What's the blue?"

I could see that Buz was glancing between the biker and the snake. I tapped him on the leg with the handle end of the snake stick. He stuck his hand backward and I placed it so he was able to grasp it without calling any attention to the movement.

Buz wormed his way closer to the snake. I elbowed my way to his side. It was hot outside, especially after riding in an air-conditioned car for over an hour. I was sweating, wishing Buz would hurry up and bag that snake so I could get a soda.

"The blue is a mineral known as chrysocolla," Buz informed the Biker. "Not quite Bisbee Blue turquoise, but it will make a nice cabinet specimen nonetheless."

As the crowd stepped back to try to get a look at the rock, Buz and I moved toward the loudly rattling and randomly striking snake.

"Hey," the biker said. "That's a timber rattler. If that thing bites you, you'll be dead in sixty seconds."

Buz quickly pinned the snake and picked it up with no trouble. I ran over to the car and grabbed a pillowcase that was once red but had faded to pink over time. To the astonishment of the crowd, Buz dropped the snake into the pillowcase and tied a tight knot.

He looked at me and quietly let me know that the snake was not injured. Then he regarded the biker who was staring at his rock. "It's a Western Diamondback Rattlesnake. Timber Rattlesnakes are not found in Arizona. I believe the closest you can find them to here would be Oklahoma and Texas."

Later I asked him why he didn't correct the biker as to the deadliness of the Timber Rattlesnake.

"It was a close call, Richard. I almost did, but that dude was so overweight and breathing so heavily, and was so afraid to throw that worthless rock, that he might have died in a few minutes from just about any snakebite. I didn't want him to feel any more secure so he could throw rocks at any other snakes."

A short time later, while Buz smoked a cigarette and I sipped from my 64-ounce Polar Pop, we drove down the highway looking for a secluded spot to release the rattlesnake.

"It's a good thing none of those guys had guns," I said. "It could have ended badly."

Bisbee in the Late 1970s

When copper mining ceased in Bisbee, Arizona, in the mid 1970s, the town went into a deep depression. Businesses languished then closed. Artists and hippies (sometimes they were one and the same) moved in. In the early 1980s galleries and quaint shops and restaurants opened. This photo shows a depressed street with nearly all stores closed. Can you see Richard standing in the shadows?

Many by-products of Bisbee copper mining are considered some of the finest in the world. High quality turquoise, some of which is well-known as "Bisbee Blue," as well as malachite, azurite, cuprite, wulfenite, aragonite, and calcite are just a few of the minerals that may be found in museum cabinets world-wide and in shops and galleries in Bisbee.

Chrysocolla, like that which appeared in the rock that biker couldn't quite bring himself to hurl at the rattlesnake, is a type of copper ore. It is pretty, but not considered valuable.

Quirky facts about Bisbee

In the movie, *Close Encounters of the Third Kind*, a character who had been abducted by aliens and returned to earth is identified as "John DeLorean, Bisbee, Arizona."

During its mining heyday in the Nineteenth Century, Bisbee was the largest city between St. Louis and San Francisco.

Bisbee is featured in both the 1957 and 2007 versions of the movie, *3:10 to Yuma*.

On one episode of the H.B.O. television series *Curb Your Enthusiasm*, Larry David discovers that he was adopted. He travels to Bisbee, Arizona (Spelled Bisby on the show) to meet his real parents.

The Cochise County seat was moved from Tombstone to Bisbee in 1929.

In the movie, *L.A. Confidential*, the character played by actress Kim Basinger is from Bisbee. At the end of the film, she says that she is moving back home to open up a dress shop, because the "Bisbee ladies need fashion too."

Bisbee has had a "hugging mayor" for at least three terms. She introduced herself to Irisse and me one time when we entered an ice cream parlor. "Hi, I'm La Verne "Williams, the hugging mayor," she told us. We each got hugged.

*"Strange things are afoot at the Circle K," is a line from the popular movie, Bill and Ted's Excellent Adventure. References to Circle K convenience stores can also be found in song lyrics by Weird Al Yankovic, Screeching Weasel, Teenage Bottlerocket, Spiderbait and OK Go.

**See *Jerry and the Duck*, Snake Hunting on the Devil's Highway

<u>Strange Arizona State Law</u>

Donkeys cannot sleep in bathtubs.

Petroglyph in the American West

CHAPTER 4
Snakes and Ass on the Oatman Road

I felt just a bit like Old Santiago, Ernest Hemmingway's character from *The Old Man and the Sea*. He had gone fishing eighty-four days without a catch. I had tried the Oatman Road twelve times, looking for snakes, not fish, and finding only a small fox, a large scorpion, and several wild burros. All my years of experience told me that this road had all the right indicators for finding snakes at night.

I call it The Oatman Road because it leads to the old west town by that name but it is actually called Boundary Cone Road, named for a huge cone-like rock formation in the area. The road begins at Arizona Highway 95, in Mohave Valley, Arizona, not far from where California, Nevada and Arizona meet. It winds about thirteen miles to the living ghost town called Oatman. To the immediate north of the intersection of Highway 95 and Boundary Cone Road is Bullhead City, Arizona. The gambling resort of Laughlin is across the river from Bullhead City on the Nevada side. Las Vegas is about 90 miles to the north.

The first time I went to Oatman I was just a teenager. I had explored much of Route 66, which at one time had gone right through the center of town and continued on to Kingman. I remember thinking even back then that it would be fun to look for snakes around there.

When first turning onto the road from Highway 95 you pass a golf course on the north side. That golf course is watered at night, and there is often water on the side of the road. It would not surprise me to find any kind of amphibian in or near those large temporary puddles, as well as garter snakes, should they occur in that range.

The road climbs gently in elevation as it heads for Oatman. At first you find lots of desert vegetation, but no cacti. After you pass the cut-off to the middle school on the right and the road that leads to the dump on the left, you begin to see boulders, which increase in number and size the farther along you go.

In the stretch of road that is bordered by boulders, I envision lyre snakes and rosy boas. In the other areas I picture rattlesnakes, lots of rattlesnakes, as well as long-nosed snakes, glossy snakes, kingsnakes, gopher snakes, and sand snakes. Banded Geckos would be on the road at night, and possibly even a gila monster. In the early mornings before it gets too hot, I predicted that horned lizards, desert iguanas, chuckwallas, leopard lizards, and collared lizards would be out sunning.

Once you begin to approach the town you need to slow down, even if you are already cruising slowly. There are wild burros all over, including in the middle of Oatman itself. Sometimes they congregate in the road, and you have to navigate around them. This is by day as well as night. When the shops are open in town, you can purchase carrots and offer them to the burros, but so many people do this each day that the animals sometimes do not have the need to feed or are not in the mood for food.

The town is scenic, historic and fun. A gunfighter group holds shootouts in the middle of Main Street every day. The burros are a bonus. All of these attractions are entertaining and interesting, but they won't get you any snakes. To get snakes, you will have to go cruise the roads at night, which is what I tried twelve times in a six week period one year with no good results.

When you've been a snake hunter for many years, you learn that there are several factors that can impact your success. Weather is one of the most important. Snakes will simply not be out on the roads if it is not warm enough. Wind can be a bitch, but Buz and I have caught hundreds of snakes in Mojave, California, and the wind is nearly always blowing there. Moon phase is important. You can clearly find more snakes in the dark than when the moon is full and the desert is all lit up. Rain can be a dynamic force when snake hunting in the desert. Rain brings all manner of critters out, including what snakes eat. It can be extremely productive to cruise desert roads during or immediately after a summer rain.

Keeping all these factors in mind I cruised the Oatman Road twelve times, from the beginning of July to the middle of August, in all moon phases and even during and after a summer monsoon had swept through. The only factor that was constant was the heat. The area triangulated by Needles, California to the south; Bullhead City, Arizona to the North; and Laughlin, Nevada across the Colorado

River to the west is one of the hottest geographies in the country. Daytime temperatures during my six weeks' stay often reached 116 degrees and only dropped to the low 100s between eight pm and midnight. This was considerably warmer than anywhere else I've ever hunted for snakes. I made up my mind I would try again in the spring.

"No Richard," Buz told me on the phone. "April is too early. We need to wait at least 'till the end of May. It would be a waste of time and gas to go now."

"But," I protested, "I'm checking the temps every day. I'm telling you, man, it's different out there on that road. Late May would probably be good, but it should also be good now. When it's good around Tucson or in southeastern Arizona during the summer monsoons, I believe it's too late on the Oatman Road. I don't see why snakes should have a shorter season somewhere else in the state. I think they come out of hibernation earlier, they mate earlier and all that on the Oatman Road. Besides, if you like the road we can go back in May."

Main Street in Oatman

Formerly a part of the famous Route 66 highway, this section is now home to daily staged gunfight shows, wild burro descendants of the original mining stock, and (alas) the ever-present T-shirt shops.

On April 20th Buz and I made our first pass together on the Oatman Road. It was just getting dark, which to me was perfect because I wanted him to see how the desert looked there, and how it changed as we approached the town. I was driving. As we cruised along the nearly deserted road I leaned forward in my seat anticipating spotting a snake at any moment. I also told Buz what I had learned about the town.

"There never would have been a town out here if it weren't for the discovery of gold in the area. That happened in 1902. From then until the 1940s Oatman and the nearby town called Gold Road were the largest producers of gold in Arizona." I swerved hard to the right but it was only a white spot in the road. Buz laughed.

"The town was first known as Vivian, because that was the name of the first company to develop the mines there, but the name was changed to Oatman in 1909 to recognize Olive Oatman who had been held first by Apaches and then by Mohave Indians. By the time she was rescued in 1857 her face had been disfigured by Mohave Indian tattoos, and Olive Oatman had not been accepted back into white society."

"Stop," Buz said softly.

"All right," I replied. "I guess I've told enough history, except..."

"No, didn't you see that snake back there?"

I stopped in the middle of the road. "No," I said. "Guess I was too busy yakking." I started to turn the car around but Buz had opened the door and hopped out with a flashlight and snake stick.

I turned the car around and aimed the headlights down the road. Buz was already on the snake and had it pinned.

"Grab the bucket with the twist-off lid," he yelled. This one's a nice Mitchell's.*

Buz was interested in taking that one home for a series of lectures he had lined up at one of the state parks, so we secured that specimen in a big painter's bucket. Then we turned the car around and headed once more toward Oatman.

We were approaching the town when I stopped to observe two burros standing on the side of the road, partially on the road and partially on the sandy shoulder.

I made a comment about how cute they were and how these were generations left over from those used for work with the mining.

"Yeah, they're cute, Richard, but did you see that Funniest Videos show where one of these cute little critters jumped the bones of that guy?"

"What?" I said.

"Yeah," kicked the crap out of him and then tried to ride his ass."

"Ain't gonna work, Buz."

"What?"

"Trying to get me paranoid. I love these creatures."

"I'm just telling you about what I saw on t.v."

"Fine, let's find some more snakes."

Approaching Oatman the road takes a turn to the left. When we made that turn I saw another rattlesnake in the sand off the road, and pulled over.

"Rattlesnake," I said, "over in the dirt just behind us. Buz was out with just his flashlight. I grabbed a hook and a sack and ran back. Buz had the big snake in his flashlight beam.

"Look, this one's a diamondback," he said, standing in front of the snake and taking the hook from me, as I hurriedly approached. Immediately the snake turned away and attempted to flee the scene. It was only when Buz pinned it that it began rattling like a snare drum, breaking the desert silence with a vengeance. We also took that snake for Buz to use as a lecture specimen.**

A few minutes later we were driving along a deserted Main Street in Oatman.

"See this hotel?" I said pointing to the Oatman Hotel as we were going by. "On March 18th in 1939, Clark Gable and Carole Lombard got married in Kingman and spent their honeymoon night in a suite right here. During the day you can go up and see the room. That brings a lot of tourists in, you know geezers like us, not kids. Most kids today don't have a clue who people like Gable and Lombard were, or Bogart, Cagney, Cary Grant, Jimmy Stewart, or Bette Davis."

"You sure are talking a lot tonight, Richard. You're usually much quieter," Buz told me.

"Maybe I figured if I just keep talking you won't try to freak me out with any tall tales or high stories. By the way, I recently found out that Clark Gable and Carole Lombard were only married for slightly less than three years. On January 16th in 1942, she was returning from a defense bond campaign in Indiana to raise money for the war effort when her plane crashed into Table Mountain, over in the Potosi Range in Nevada. She and twenty-one others died in that crash. That's it. I'm shutting up."

Beyond the town the road climbs and winds into a mountain pass that leads to Kingman. There is a mine or mill on the left side, then nothing but a treacherous tangle of twists and turns. We stopped to look at a trio of wild burros huddled together in a turnout. Then, just a bit further we found another Mitchell's. This one we caught and released. We turned around at the next opportunity and headed back the way we came. Only this time, we took the cutoff toward Golden Shores (also called Topock). The geology is different along a stretch of this road, with fewer boulders, more sand and almost no vehicular traffic.

It wasn't long before we stopped for a sidewinder, which we photographed but did not pick up, other than to scoot it off the road. Shortly after that we stopped for what we assumed was a glossy snake, by the way it reflected pure white, but by the time we got out of the car it had vanished without a trace.

Buz and I just looked at each other and shook our heads in shame for letting a live one get away.

Next came a leaf-nosed snake. This tiny docile snake was easy to pick up, photograph and release. During this process we took a break and drank some coffee from a thermos. We looked up at the magnificent desert starry sky, away from electrified civilization. The constellations were clear and bright. It reminded me that it's a good idea to look up from time to time, even when you are spending most of your time looking down.

Back on the Oatman Road in amongst the boulders, we chased a patch-nosed snake, both Buz and I getting our hands on it but the wiry little snake was hot and it slipped through our fingers and escaped into a creosote bush.

"That's two that got away in one night, Buz. We must be getting old. This never would have happened when we were young. And what's a patch-nosed snake doing out at night anyway. I've never seen that before, have you?"

"No," he replied. "You told me on the phone that it's different out here. I guess you were right."

And then he looked at me and added, "It's about damn time!"

*Actually called Southwestern Speckled Rattlesnake (Crotalus mitchelli), experienced snake hunters would never say, "Look at the beautiful Southwestern Speckled Rattlesnake I just caught." Nor would they say, "I just picked up a Western Diamondback Rattlesnake and a Mojave Rattlenake." There are two kinds of diamondbacks: Western and Eastern. Any experienced snake hunter would know which one is in the geographic zone being hunted. Informally, the conversation would go like this: "Just picked up an Atrox and a Scut (pronounced scoot)," or "Just picked up a diamondback and a Mojave." Formal language requires scientific names: "Just picked up one Crotalus atrox and one Crotalus scutulatus".

**At the end of the summer, both rattlesnakes and some non-venomous ones from subsequent trips were brought back and released.

Mule Tales

What is the difference between a burro, a donkey, an ass and a mule? What are Jacks and Jennies?

A burro, donkey and ass are three different names for the same animal. A Jack is a male and a Jenny is a female of any of these.

A mule is different. It is a hybrid from a female horse (mare) and a donkey (sire).

Horse: Equus caballus

Donkey, Burro, Ass: Equus asinus

The mule, being a cross between two different, although very similar, creatures, is sterile.

Stubborn as a mule? Here is what Harry S. Truman, 33rd President of the United States, had to say about that:

"My favorite animal is the mule. He has more horse sense than a horse. He knows when to stop eating - - and he knows when to stop working."

Wild Burros

Wild burros looking for a handout on the main drag through Oatman.

Images courtesy of Keith Davis

A Young Burro

This young burro is taking a nap in the shade, just off Oatman's Main Street in a public area.

An Older Burro

This older burro munches on hay pellets. At night it will wander out of town into the surrounding desert.

CHAPTER 5
A Mysterious Disappearance

Not far from the premier snake hunting grounds around Willcox and the Chiricahua Mountains can be found the remains of an interesting stagecoach station along the historic Butterfield Trail. Buz had been there before. He wanted to show it to me and give me (I guess) a history lesson on this hot, humid day in August.

We were staying at the Best Western Plaza Inn in Willcox, which was newly built at that time. After a breakfast at the truck stop directly next door to the motel of eggs, potatoes and toast, along with a side of their excellent homemade salsa, we were gastronomically satisfied and ready to embark on a late morning adventure. The first thing Buz told me was that we would possibly find some reptiles out there.

"In this heat?" I asked. "What, racers and whipsnakes?"

"Yes," he said. "But also lizards. Last time I was there a few years back I found three big spiny lizards, I think they were Clarks. Two males were trying to court one female. The males had beautiful iridescent colors. But you can also find horned lizards in the sand, and you never know what might be out on the rocks or around the structures. That's the kind of place you can find gopher snakes during the day, and keep a lookout for big diamondbacks in the ruins, you know in the shade. And I wouldn't totally rule out the possibility of a gila monster walking around out there."

I smiled at that. "Diamondbacks and gila monsters. Yeah that's what I'm talking about."

Buz cruised the red and white International Scout onto Highway 10 going west. Then he began telling me about the stagecoach station.

"Around 1858, before the Civil War and before the railroad, a stage line called the Butterfield Overland built a trail from St. Louis to San Francisco. I think they also had some southern routes, but

I'm talking about St. Louis to San Francisco. When the surveyors got out here, they knew of a nearby water hole."

We passed the Highway 666 South off ramp. The night before, we had made a snake hunting run on that road, from the exit we just passed where it meets Highway 10 all the way down to the small, agricultural town of Elfrida. Then we worked our way back up here and did it again. Between seven forty-five p.m. and one a.m. we found alive: one Sonora Kingsnake, one Western Diamondback Rattlesnake, one Mojave Rattlesnake, one Painted Desert Glossy Snake, one Spotted Night Snake, and two Sonora Gopher Snakes. Dead on road included one Mexican Hognose Snake, two Mojave Rattlesnakes, one Diamondback Rattlesnake and two Sonora Gopher Snakes.

Just past the Highway 666 off ramp Buz continued his story.

"The surveyors knew of this water hole, but because Cochise's Stronghold is right out there (he hand-cranked the window open, extended his arm straight out and pointed south), and it was well-known that the Apaches relied on that water hole, they decided to build the stage station one mile west of it, so as not to antagonize the Indians. Turns out they had problems that didn't include Indians, at least at first.

"While they were building the station, three employees were

Dragoon Springs Stage Station

A beautiful setting for a stage station in the Dragoon Mountains. As part of the Butterfield Overland Stage route from St. Louis to San Francisco, this station was built in 1858 as part of their mail and passenger route. In those days there was a natural spring nearby. During construction unthinkable acts of savagery occurred there as three employees were robbed and clubbed to death by Mexican laborers. The foreman had his left arm hacked off by an axe. Later, four Confederate soldiers were killed by Apaches. Today there is peace among the ruins as warm breezes fill the air with promise for those looking for reptiles, western history, or just gorgeous scenery.

attacked in their sleep and beaten to death with heavy mining tools by Mexican laborers. The motive was robbery. The foreman, whose name was Silas St. John, had his left arm hacked off by an axe, but still managed to fight back. Eventually he chased off the attackers. Not only that, but he somehow also fought off the buzzards and wolves that were going after his dead friends.

"He had lost a lot of blood, but three days later help arrived and a doctor was sent for. Did you notice that I said wolves? Well they aren't here now, but wolves sure as hell were here then, just as jaguars were over in the Chiricahua Mountains back then."

At this point we took the Dragoon Road exit and drove toward the tiny town of Dragoon. Naturally I kept eye-sweeping the road in case any critters might be out.

"The three Overland employees are buried out here next to what's left of the station. So are four Confederate soldiers who were killed by the Apaches. This was four years later, just after the famous Bascom Affair, where Cochise, who was agreeing to peace terms, was betrayed and deceived by the U.S. Army. After that, any soldier was seen as a threat to the Apaches."

Buz braked slightly as we swerved over a rut in the road. "I bet you didn't know that Arizona, which was only a part of New Mexico territory at that time, had seceded from the Union in 1861."

"Why would they do this?" I asked.

"Two reasons. One was to support the Confederacy and its values as related to the Civil War. The other was to break from New Mexico and gain separate territorial status for Arizona.

"Captain Sherod Hunter's regiment of Rangers were rounding up stray cattle in the vicinity of the watering hole one mile east of what remained of the stage station. When they entered a narrow box canyon, they were ambushed by Cochise and about a hundred Apache warriors. Most of the Rangers escaped, but four did not. They're buried near the men who were building the station. So there are a total of seven graves out here."

We crossed over the railroad tracks and followed the road as it curved to the right. Then we turned right onto a dirt road and followed it as it arched around the northernmost section of the Dragoon Mountains. A few seconds later Buz hit the brakes, and I flew out of the car to chase a horned lizard into a mesquite bush a few feet off the road.

"That's perfect," I remarked to Buz as I climbed back in the car empty-handed. "A lizard with spikes runs into a bush with thorns. Guess there's a thousand things that can stick you out here. By the way, don't you love it when people call them horny toads?"

Buz drove along the dirt road and we both watched for anything moving in the dirt ahead of us.

"What is that I always hear you tell people at lectures when they say something about horny toads?"

Buz looked my way. "I tell them I don't know anything about their sex lives, and they're not toads."

Of course I knew what he would say. I must have heard it a couple dozen times, but I still had to smile hearing him say it again as we drove along this deserted dirt road with a great view of the mountains, the opportunity to find more critters, and an interesting historical site just ahead. We passed a Forest Service sign reading "Butterfield Overland Stage Station," and Buz pulled over and parked.

Outside there was a warm breeze. I spotted the piled rock remains of the stage station and I crossed a narrow gully on a crude cement bridge to walk in that direction. Buz was hanging back and snapping a few pictures. Soon I was standing inside one of the rooms of the old ruins. I closed my eyes to the bright sunlight and tried to imagine the horror that occurred here. A disturbingly ugly dude in my mind was swinging an axe while another specter with war paint was shooting arrows. Quickly I opened my eyes. I wanted to tell Buz about it, but when I turned around, he was gone.

"Hey Buz!" I yelled, the sound breaking through the deserted wall of quiet.

I waited but there was no answer. I slowly turned panoramically and saw lots of beautiful scenery, but no Buz. My first thought was that he was messing with me. He would do that from time to time, and then laugh and tell a wild story about why he did it. But then I shivered with an eerie feeling that something was wrong. I moved around a bit and glanced over my shoulder so I might get a quick glimpse of any axe-wielding or arrow-shooting phantoms still lurking around. But no, I thought. If any supernatural elements were here, it would more likely be the ectoplasm of those who had been killed.

Buz was nowhere in sight and I was starting to get concerned. I no longer felt he was messing with me. Now I wondered if he had been abducted by aliens in broad daylight. But wouldn't I have seen something, heard some sounds?

I began walking toward the car. When I reached what was left of the stage station's last rock wall I looked down and there was Buz, completely sprawled out on the rocky ground in a twisted position.

"Man, what happened? Are you hacked up or shot?"

"No, Richard, But I figured you'd think of something like that. I was looking at that Collared Lizard over there, and I tripped over a rock and went down hard. Man I hope I didn't break my ankle."

"Oh geez, man, did you hit your head, are you hurt anywhere else, do you think you can walk? What Collared Lizard?"

Buz slowly raised himself to a sitting position and leaned against the rock wall he had tripped over, groaning and trying to feel his ankle through his rough-out cowboy boot. I preferred to keep standing due to what I assessed to be a near perfect arachnid ecosystem, potentially containing legions of small but toxic creatures, disturbed by Buz's fall, eager to emerge from hiding, seeking revenge.

"No, yes, I don't know and over there." Buz pointed to the rock wall, perpendicular and about twenty feet from where he was

Buz Went Down Hard While Watching a Collared Lizard

I wondered if he had been attacked by phantoms of the stage station, or shot by pissed-off Apache ghosts. Buz hoped that his ankle wasn't broken. I snapped a picture before lifting him up.

Now I Wondered...

Now I wondered if he had been abducted by aliens in broad daylight.
(Alien Fresh Jerky, Baker, California)

leaning. And there was a small Collared Lizard, green and speckled with its head tilted in our direction, definitely aware of our presence.

"Wow, look at that," I said.

"I did," Buz replied, "That's why I was doing the backstroke in all this dirt and rocks. I was trying to take a picture, walking forward slowly, looking through the lens."

"Did you get a picture before you went down?"

"I snapped something, but it will probably be sky."

"All right," I said, "You rest for a minute. Why don't you give me the camera?"

Before Buz handed me the camera, I spotted some commotion with the Collared Lizard. I thought it might have grabbed one of the many dragonflies circling around. But that wasn't it at all.

"Did you see that?" Buz asked.

"I saw something. What was it and where did the Collared go?"

Buz immediately tried to stand up. "A big ol' bilineatus* came up and collared the Collared."

"Holy crap." I reached down and offered both hands to Buz, but he stubbornly ignored them and pushed himself up, moaning when the weight of his body engaged his twisted ankle.

I took the camera and left Buz standing there, slowly attempting to walk. I almost reached the spot on the rock wall where the lizard had been ambushed, when the quiet was shattered by a thunderous buzzing. Just in front of me, coiled in the shade, its black tongue flicking out and in, was a large diamondback rattlesnake. I swiped at a curious dragonfly that invaded my comfort zone, and turned to look at Buz.

"I heard it, Richard," he said and limped along toward me. I snapped a couple of pictures. Then I realized that we didn't bring a hook.

"You keep an eye on him," I said. "I'm gonna run to the car and grab a stick."

Buz was grimacing with ankle pain as he stood and regarded the diamondback, its black and white "coon tail" pointing up, its rattles shaking in a blur.

By the time I got back with the hook, Buz was holding the snake and smiling.

"Why didn't you wait?" I asked.

"It started moving away," he said.

"What'd you pin it with?"

"My boot."

"Geez, man, the one with your good ankle or your mangled one?"

"Had to be the mangled one, 'cause I couldn't get the balance using the good one."

"That's a beautiful snake, man; nice bright colors, must have recently shed. Now it's probably looking for a meal."

Buz turned the thick-bodied, four-footer this way and that to get a good look all around. "I don't think this one was looking for anything. I think he was just basking in the shade enjoying his day until you came along and disturbed him."

"I came along?"

Buz Posing with Diamondback Rattlesnake

In the late 1970s there were no digital cameras or cell phone cameras, where you can check the quality of your pictures before moving on. Our pictures of Buz holding the Diamondback that day did not come out well enough to include in this story, so we went back the following year. Here he is holding a different, but similar Diamondback caught in approximately the same location. The snake was released, unharmed, before leaving the area.

"Yeah, remember you were gonna take pictures of the Collared, then that whipsnake came up and grabbed it. You were going to check it out when this one advised you not to step on him. Then I came over and here we are. Take a couple of pictures. Then you can hold it and we'll let it go."

We did all that, and I must say that holding a large, beautiful and potentially dangerous snake like that can really boost the spirits of any reptile enthusiast.

After releasing the rattlesnake we searched the area for the whipsnake and the Collared Lizard. Whipsnakes are not constrictors like kingsnakes and gopher snakes, so they don't throw coils around their prey. Rather they must over-power their prey and either pin them to something or just swallow them alive. We searched along the rock walls and in and around all the nearby bushes, but no luck.

"That could have made some good pictures," Buz observed.

I agreed and for a moment regretted splitting up and going for the snake hook we didn't even use. "How's your ankle?" I asked.

"Could be better, but I guess I'll live."

I walked and Buz limped toward seven mounds of dirt near the opposite end of the ruins. These were the graves of the seven men who died around here. Only one was marked, that of Sergeant Samuel Ford, a member of Governor John R. Baylor's Company A Arizona Rangers Dragoons. When we reached the graves, we both removed our hats and observed a moment of silence.

Then we put our hats back on and headed slowly toward the car.

"So why did they use rocks to build this station," I asked as we walked.

"Not everything around these parts is made of adobe," Buz replied. "And another thing. There was a major earthquake down in Bavispe,

Collared Lizard on Petrified Rock

Not the unfortunate lizard that was snatched by the whipsnake, this one was at home sitting on a petrified log in the Petrified Forest National Park.

Mexico that killed a lot of people down there and did damage for hundreds of miles. You know that nearby spring down here I was telling you about? The earthquake closed it up. So Dragoon Springs has been without a spring since that earthquake in 1887. This is the third time I've been here, and each time, including right now, I can't help thinking about old Silas St. John and how he got his arm hacked off and still managed to fight off his attackers and guard his friends' bodies for three days and nights until help arrived. I tell you, Richard, this is sacred ground for acts of bravery in Arizona history."

"Yes," I said. "And it's not too shabby ground for current natural history. Thanks for bringing me here."

*Bilineatus is Masticophis bilineatus, the scientific name of the Sonora Whipsnake, a long, slender bluish green snake, whose bright colors fade out about midway down the body. I think Buz was too stunned to think of the common name, what with his painful ankle and the suddenness of this rarely witnessed event.

Crazy Arizona County Law – Mohave County

"Anyone caught stealing soap must wash himself with it until it is all used up."

Strange Arizona City Law – Hayden

If you bother the cottontails or the Bullfrogs you will be fined.

CHAPTER 6
The Strange Winged Monster

On this particular moonless night our snake hunt began just before 8:00 p.m. when the daylight had not yet completely deserted the sky. Driving at thirty-five mph, we first made a loop from Willcox, Arizona on Highway 186 past the still-living ghost town of Dos Cabezas to the entrance of the Chiricahua National Monument, then continued on Highway 181 until we reached 666, the *Devil's Highway*.

This loop can be extremely productive for snake collectors. Several times in the past we had found eight to twelve snakes on one pass. And the species are varied. Three different rattlesnakes live in proximity of each other: Western Diamondbacks, Mojaves and Black-tailed. Sonora Gopher snakes are common, as are Long-nosed snakes and Painted Desert Glossies. Sonora Kingsnakes (now called Desert Kingsnakes) are regularly found on these roads, and Mexican Hognose snakes are often out during or after a rain. Black-necked and Checkered Garters are out both day and night, and occasionally one can find Western Ground Snakes, Spotted Night Snakes and Western Coral Snakes.

That night we found nothing. Not even a road kill. Or a toad. Nada!

This can happen to hunters, fishermen, bird watchers, anyone. It's frustrating, because you know the critters are there, you've only got a limited amount of time and at least a few snakes should have been out. When experienced fishermen get skunked, what do they do? They move to a different area. That's what we did that night.

We drove from the ranchland of Willcox, elevation 4168; to the desert of Tombstone, elevation 4540; and continued our snake hunt on the Charleston Road and a number of small roads between Tombstone and Sierra Vista.

At 11:20 we stopped for an extra large Western Diamondback Rattlesnake basking on the warm asphalt, its "coon" tail raised like a black and white flag. It was a one-in-a-thousand, premium quality specimen, fat and feisty enough to protest being hooked much more so than most Western Diamondbacks. Buz pinned it, picked it up, and we stretched it out on the road and measured it. That was no easy task, as it thrashed about, all the time buzzing forcefully. Finally we got the measurement: fifty-eight inches. It reminded me of a Western Diamondback I used to observe at the Los Angeles Zoo when I was a teenager, which was longer and even bulkier than this one; however this snake was wild and had no zookeepers tossing food in its vicinity.

Buz and I looked at each other. I knew what he was thinking, and I said, "We really should release it." He didn't say anything for a minute, just stood there holding it and admiring it. Then he nodded.

"Hold open a sack. We'll take pictures in the morning. Then we'll release it."

That woke us up. We were all smiles as we stood leaning against the car, drinking thermos coffee from previously used Styrofoam cups, listening to the beautiful Connie Francis sing *Everybody's Somebody's Fool* from the car's cassette tape player (this was after 8-track and years before compact discs were commercially available).

Most people are perfectly content to live their lives without snakes. To these folks it is difficult to explain the absolute thrill of catching and handling any rattlesnake. Maybe it's the danger, knowing that you could be a split second too slow, or the snake could twist out of a hold. Although not at all common, several experienced herpetologists and non-professional handlers had been messed up, even killed by rattlesnakes. But to catch-hold-measure-and-release a true paradigm... man, it's like reaching-the-top-of-the mountain-exciting.

We were still leaning against the car consuming coffee when I said, "Damn, that is one trophy diamondback. I don't believe I'll ever forget it."

Buz nodded in agreement.

"Man, you're not saying much tonight. That's not like you."

"Well," he said, "I was just thinking. You know about the strange winged creature that was found right out here in the desert by two cowboys in 1890, don't you?"

"No," I said. "What kind of creature?"

"It was said to have a head like an alligator, smooth, featherless wings, only two legs and smooth skin."

"Sounds like a bat with an alligator head."

"Not really. A bat is more like a mouse with wings. The total length of this creature was over ninety feet and the width was around four feet."

"What?"

"Yeah, and the wingspan was about a hundred and sixty feet."

"Okay," I said chuckling, "Where did you hear this?"

"From an old dude in a rock shop."

I chuckled again. "You mean the one that told you about the vein of silver in the bat cave over in Gunsight?"*

"Yes, and he was right about the vein of silver. Remember, we went in and chipped some off?"

"Yeah, but I also remember he said there weren't any bats."

"No, I told you that because I knew you wouldn't go in there if you knew the truth."

"And isn't he the same one that told you about the flying saucer that crash-landed out in Tucson?"**

"Absolutely."

"And now that I think of it, didn't he also tell you about all those people being shot over in Rodeo? Something about a masked man riding his horse up the steps of the saloon and blowing hippies out of their bar stools?"***

"Yes, and he was right about all of that. There's plenty of evidence."

I poured each of us another half cup of coffee, and looked all around as nonchalantly as possible.

"Okay, describe this old dude."

Buz stared at me for a moment, sipped some coffee and said, "About seventy-five to eighty years old, grizzled white hair and beard, dressed in miner's clothes, had one good eye and one wandering eye with a long scar over and under it, where it was hard to look at him when you were talking. The good eye had a wild look to it, but he spoke from experience and knew everything about rocks, minerals and the general history of this region. I think he said his eye got messed up in a blasting accident. I bought a few rock specimens from him before I knew you, and got a lot of useful information each time I went there."

"What kind of rock specimens did you get from him?"

"Oh, I got some pyrite, calcite, azurite and malachite mainly. A few others."

"All right, what did he say about this so-called monster?"

"He said two ranchers came across this creature as they were riding home from the Huachuca (wah-chew-ka) Mountains. They thought it was exhausted from a long flight because it was only able to fly short distances at a time. When they got over the shock of seeing such a thing, they pursued it for a mile or two, but it was difficult because their horses were freaked out and all. When they got close enough, they wounded it with their rifles. Now the beast was really pissed and went after the men, but they kept shooting it and it finally rolled over dead. When they mustered up the courage, the cowboys approached it, made some measurements and cut off a chunk of one of the wings, which they brought into Tombstone and showed it around."

Buz paused and drank some coffee.

"Anything else?" I asked.

"Well they said the eyes were the size of dinner plates and the wings were composed of a thick, nearly transparent membrane. Let's see, it had the head like an alligator, full of sharp teeth and smooth skin with no fur. And the old miner dude told me that a posse of men brought the creature back to town and nailed it across the entire width of a barn." At this point Buz paused. He turned and looked me in the eyes and said, "What do you think?"

I chuckled, then I said, "I think the old miner dude was having fun with you the same way the corpulent owner of that desert café played me for a dufus.**** There's a tradition of that in the west and..."

"No," Buz interrupted. "The old miner dude was telling the truth."

"Come on, man. How do you know?"

"First of all, he was right about everything else. You saw the bullet holes in the saloon out in Rodeo. You've got a souvenir piece of silver from the vein in the bat cave. And every UFO enthusiast who knows his ass from a hole in the ground knows about the flying saucer that crashed in Tucson. Not only that but it stands to reason that these creatures were seen from time to time because why else would Zunis and other Indians use thunderbirds as religious and cultural symbols? Think about it. And if all that's not enough for you, the old miner dude told me that the Tombstone Epitaph newspaper published an article at the time of the event."

"That's sexist, man."

"What is?"

"Are all UFO enthusiasts male? You gotta figure that at least some of them are female. You said every UFO enthusiast who knows *his* ass from a hole in the ground."

"Come on, Richard. When have you ever known me to be politically correct? Besides, what if all the E.T.s are females and all they want is males?"

I searched Buz's face to see if I could get a reading, but his expression was blank. On purpose I supposed. Finally I concluded that he was messing around so I said, "Yeah, right, fine, but if I decide to write about this one day, you're good with your statement as is?"

"I don't see anything wrong with it."

"Okay, I'm making a note of your exact "his ass," politically incorrect quote. And about the flying monster, I've got to say that you make an all right case, but I still don't believe it."

"Well, what if we can find the article, would you believe it then?"

"I don't think it exists, so I seriously doubt we can find it.'

Buz stared at me. "I'll bet you a sit-down dinner, two pitchers of beer and a tank of gas that we find it."

"Okay", I said, "you've got yourself a wager, Mr. Gambling Man."

We got in the car and continued looking for more giant rattlesnakes, pterodactyls and smaller, less dangerous creatures. After a few minutes Buz said, "I'm pretty sure I know what the creature was doing here in 1890."

"Oh yeah," I said. "What?"

"It was laying eggs."

We were unable to find the article in question anywhere in Tombstone that year. However, the following summer we headed directly from Southern California to the Arizona Historical Society in Tucson. Looking through their collection of original *Tombstone Epitaph* newspapers, we actually found the article in question. It appeared in the April 26, 1890 edition. Here it is in its entirety:

A Strange Winged Monster Discovered and Killed on the Huachuca Desert.

A winged monster, resembling a huge alligator with an extremely elongated tail and an immense pair of wings, was found on the desert between the Whetstone and Huachuca mountains last Sunday by two ranchers who were returning home from the Huachu-cas. The creature was evidently greatly exhausted by a long flight and when discovered was able to fly but a short distance at a time. After the first shock of wild amazement as it passed the two men, who were on horseback and armed with Winchester rifles, re-gained sufficient courage to pursue the monster and after an exciting chase of several miles succeeded in getting near enough to open fire with their rifles and wounding it. The creature then turned on the men but owing to its exhausted condition they were able to keep out of its way and after a few well di-rected shots the monster partly rolled over and re-mained motionless. The men cautiously approached, their horses snorting in terror and found that the monster was dead. They then proceeded to make an examination and found that it measured about nine-ty-two feet in length and the greatest diameter was about fifty inches. The monster had only two feet, these being situated a short distance in front of where the wings were joined to the body. The head, as near as they could judge, was about eight feet long, the jaws being thickly set with strong, sharp teeth. Its eyes were as large as a dinner plate and protruded about half way from the head. They had some difficulty in measuring the wings as they were partly folded under the body, but finally got one straightened out sufficiently to get a measurement of seventy-eight feet, making the total length from tip to tip about 160 feet. The wings were composed of a thick and nearly transparent membrane and were devoid of feathers or hair, as was the entire body. The skin of the body was comparatively smooth and easily punctured by a bullet. The men cut off a small portion of the tip of one wing and took it home with them. Late last night one of them arrived in this city for supplies and to make the necessary preparations to skin the creature, when the hide will be sent east for examination by the eminent scientists of the day. The finder returned early this morning accompanied by several prominent men who will endeavor to bring the strange creature back to the city before it is mutilated."*****

When we found that article, I knew that once again I had been outsmarted by Buz. I had no choice but to pay the bet. I bought us an early steak dinner at a coffee shop in Tucson. Before we went snake hunting that night, I filled the car with gas. Of all the points of the bet this pissed me off the most because gas had gone up to the ridiculous price of fifty-seven point nine cents a gallon. We wondered how long we would be able to go on these trips with gas prices so high.

We caught and released several nice snakes, and stopped to take a few dead ones off the Ajo Road that night. Each time I got out of the car I looked up as well as down. I couldn't stop thinking about Buz's hypothesis that the strange winged monster had been out there laying eggs before being massacred by those cowboys.

The only snakes we kept were a twenty-two inch Spotted Night Snake that Buz really loved and an eighteen inch Yuma Kingsnake that seemed real calm and content in my hand. We quit early, about 12:30 a.m. and found a dive around the corner from our motel where I completed paying off my lost bet by procuring two pitchers of beer.

As we polished off the second pitcher, Buz wiped his mouth with his sleeve and looked at me. "You wanna know what I really think the winged monster was?"

Strange Winged Monster Pictograph

This ancient pictograph, believed to have been created between 500 and 1300 A.D., adorns the walls of Black Dragon Wash in San Rafael County, Utah. A pictograph is a drawing on rock, as opposed to a petroglyph, which is a carving on rock. Could the subject of this pictograph be related to the creature that was reportedly killed near Tombstone, Arizona? Photo courtesy of Climb-Utah.com

I brought up a tumultuous belch, giggled a little from the beer effects, and said, "I'm all ears."

"I've been thinking about it all night. I'm pretty sure the so-called winged monster is a Mayan Quatzequatel."

"A quatze-whatle?" I laughed.

Buz looked serious. "A Quatzequatel is a Mayan God. I've seen drawings of it running around on the Mayan pyramids. It looks just like a feathered serpent. The Mayans believed they were visited by Gods from outer space. And they believed that this Quatzequatel, this feathered serpent, get that, *serpent*, comes to earth for the purpose of teaching peace to the people. Isn't it just like mankind to react by blowing it away, like those cowboys did to that poor flying reptile God when they saw it flying by?"

*See *Bats in the Silver Mine*, <u>Snake Hunting on the Devil's Highway</u>
**See *Weather Factors*, <u>Snake Hunting on the Devil's Highway</u>
***See *Rodeo*, <u>Snake Hunting on the Devil's Highway</u>
****See The *Sunizona Café*, <u>Snake Hunting on the Devil's Highway</u>
*****From *The Tombstone Epitaph*, April 26, 1890, page 3

Strange Arizona City Law - Globe

*Cards may not be played in the street
with a Native American.*

Petroglyph of a Lizard - Dinosaur National Monument

CHAPTER 7
Serpent's Tooth

In the early part of the 70s when Buz and I were next door neighbors, we often did not have to travel very far for adventures or trouble. This escapade occurred on a Saturday afternoon in the summer. It began with a knock on my door.

I opened the door and saw Buz standing there with a pillowcase.

"Hey, Buz, whatcha got there?"

"Oh, I don't know. The dude said it was a Crook's tree boa."

"I've heard of a Cook's tree boa, but not a Crook's." Both of us were more interested in domestic reptiles, which we knew from the field and from keeping captives. Neither of us were experts on exotics. We were far from ignorant about them, but most of our knowledge came from books, zoos and from talking to curators and old time hobbyists.

"I *do* know that if it's a Cook's, they resemble Emerald tree boas and have big heads and lots of long, sharp teeth. What did you have to give for this one?"

"Nothing. The guy wanted to get rid of it. It's very thin. He says it won't eat."

"Well, let's take a look at it."

We walked over to Buz's garage, where he had a ghetto blaster playing a song by the Eagles.

A tall cage was already set up with fresh newspapers on the floor (I guess in case the snake decided to read about current events), a large crock bowl with clean water for drinking or soaking, and most importantly for this species, a thick driftwood branch for climbing and hanging on.

Buz untied the knot in the pillowcase and reached in. Instantly the snake threw coils around his good left arm. The right arm still has big stainless steel plates and screws in it from being crushed in a terrible motorcycle accident. To this day he still has no feeling in his right arm as, in addition to the bones being crushed, the brachial and radial nerves were severed.

"Oh no," he said as the snake squeezed tighter. "I can't use my other hand to unwind him. I need help."

"Sure," I said. "Just hold onto that big head."

It was a hot summer day. We were standing in Buz's garage, and sweating. I began unwinding the snake, which did appear to be a Cook's tree boa, as I had seen before, but not up close and personal. Today they call them Amazon Tree Boas, but back then they were only known as Cook's.

When I unwrapped the last coil from Buz's arm, the snake somehow squirmed out of his sweaty grip and instantly nailed me on the middle finger of my left hand. I yelled out as it felt like the tip of my finger had just been bitten off. It looked that way, too, as it was bleeding heavily onto the garage floor. Buz managed to maneuver the snake into the cage and I headed home for some first aid.

At home I held my left hand over the kitchen sink and watched the bottom of the basin turn red, as water from the faucet hit my finger and mixed with lots of blood. Then Irisse wandered in, observed me turning the white sink red and asked what happened.

"Oh, I just got bit by a Cook's. . ."

"What?" she interrupted.

I turned to her and I could see she was thinking that I just got into a fight with a chef or something, so I tried to tell her that it was just a bite from a snake that spends most of its time in trees and needs big sharp teeth for grabbing birds through their feathers and all, but she didn't hear me because she was laughing so hard.

She kept laughing as she went for some bandages. I was still standing over the sink, not thinking anything was very funny. When she returned, she wrapped my finger in a towel and applied pressure.

"What was so funny?" I asked.

"Oh, I thought you said you mixed it up with some cooks. Then I realized it had to be some kind of reptile."

"Where do you think I was," I asked her, "that I could have been attacked by cooks?"

"Next door with Buz," she said. "Nothing would surprise me when you two get together.

I removed the bandages the following day, but every time the tip of my finger touched anything, a severe pain shot all the way up my arm, like a shockwave. This went on for over six months, and the wound never healed completely.

As anyone who has ever injured a finger: broke it, cut it or otherwise messed it up knows, you don't fully appreciate the importance of each digit until you can't use it.

Soon I got in the habit of squeezing my finger as hard as I could. I don't know why I did this, but I caught myself doing it all the time. I'm not sure if it actually relieved the pain. If it did, it was only momentary.

Months later, while waiting to pick up one of my kids somewhere, I was sitting in the car squeezing the crap out of my finger when I noticed something white start coming out of the wound. At first I thought it was some kind of liquid, and my stomach dropped. But I kept squeezing anyway until the white was all the way out. Then I noticed it was a tiny white curved tooth, about an eighth of an inch long.

I sat in the car and wondered why I never thought of that. Then I figured I probably did think of it unconsciously, because why else would I have been squeezing, squeezing, all the time squeezing?

The pain went away immediately and the wound finally healed. It's probably a wives' tale, but I've heard it said that something like that could get in your bloodstream, go straight to the heart and kill you. I'm paranoid enough to wonder if that's true, and if that had happened to me and I had dropped dead from a heart attack, would anybody ever find out the exact cause of death? If they thought the cause of death was a heart attack, why would they look for a tiny white tooth?

And I thought if that had happened, it would have meant that I was killed from the bite of a skinny, non-venomous boa.

Before Buz had brought it home, the Cook's Tree Boa hadn't eaten for many months. On close examination it was easy to see why. It had impacted eye shields. The previous owner had paid no attention to proper husbandry for snakes of the rain forest, such as misting the cage or using a humidifier.

When we were able to properly remove the layers of the old eye shields that hadn't come off with the rest of the shedded skins, the snake was willing and able to eat. It lived for several years, and neither of us were ever bitten by it again.

Odd Arizona State Law

Hunting camels is prohibited.

CHAPTER 8
The Most Fun

When people learn of my interest in snakes, they usually ask questions. What is the biggest snake you have caught or kept? Which is the most dangerous? (Sometimes they say *poisonous*, and that instigates a lecture from me about the difference between poison and venom.) Have you ever been bitten? That's when I show them all the scars on both hands. Yes, I tell them. Several times, but not by anything venomous, except for the one time a sidewinder scratched me with one fang when I was attempting to pose it for a photo.* Fortunately it was only a scratch, and a dry one at that, I tell them. No venom. If they seem interested I explain the dry bite process.

Although there are always questions, no one has ever asked this one: What was the most fun you ever had with a snake. As I have been anxious to answer that for a long time now, I will do so here.

It happened back in the 70s when I was teaching school. All my colleagues and students knew of my interest in snakes, as I sometimes brought in specimens for the herpetology class and the herpetology club. Most of the other teachers weren't thrilled with the concept, and just stayed away. As I was leaving the faculty lounge one spring day, one of my fellow teachers approached.

"I don't know if you'd be interested," she said, "but my friend is a filmmaker, and she is looking for a big snake to use in a movie."

"A movie? That sounds interesting," I said. "Are there any opportunities for fame or fortune?"

She smiled and chuckled. "I doubt it," she said, "But I have a feeling you'll have a lot of fun."

"Why?" I asked. "What kind of movie is it?"

"Well," she said. "My friend makes underground movies. In fact she's very good at her art, and quite well-known."

"Underground? Why does she need a snake?"

"All I know is they are filming a scene in the woods with a naked woman dancing around with a big snake."

I must admit that this caught my immediate attention. "Okay," I said. "Sign me up."

"All right," said my colleague with a giggle and a smile. "Are you available this Saturday? She wants to film on Saturday."

"Yeah, I can do it Saturday, but I'll have to bring my co-snake wrangler."

When we arrived at the filmmaker's house at a remote location in Tujunga, California, I looked around at the woodsy surroundings.

I told Buz, "This is too funny. I used to live only a few blocks from here. Yeah, a lot of asthmatics come up here because the air is dryer and less polluted than just about anywhere else in Southern California. My brother had asthma back then so we moved up here for a couple years. I hated this place, except for chasing lizards in the empty fields and looking for snakes up in the canyons. I never knew this kind of secluded residential area existed."

Buz untied the large cloth bag and removed the large Columbian Red-tailed Boa Constrictor. He looked around. "Yeah, it is an interesting-looking house. Let's go in."

We were met at the door by a dark-haired woman in her mid-forties, wearing jeans, a blue work shirt, a dangling southwestern necklace and bracelet, and sandals. She had a definite twinkle in her eyes, which, like a camera, panned from me to Buz to the boa constrictor.

"I'm Chick," she said. "Thank you for coming here today and bringing that wonderful snake."

I introduced myself and Buz and explained that the two of us had several large snakes, but this particular one was by far the tamest and the calmest, and by virtue of just recently shedding its skin was undoubtedly the prettiest.

I also told her that although Chick was an unusual name, my own father had grown up in Brooklyn, where every kid in that generation had a nickname. His real name was Charles, but he was known all his life as Chick.

"Mildred," she replied with a warm smile.

I started to tell her about having lived around there, but a man walked into the hallway and stuck out his hand.

"What's this about names?" he said. "Mine's Marty, but I use Neon Park with my art. Whoa, that's a big snake."

I was immediately impressed with the warmth and enthusiasm of, what I found out to be, this married couple. Marty also had that rare quality, the gleam in the eye, that illuminates s genuine friendliness, and above all else, a lust for life.

Chick showed us through the house, the inside of which I do not remember after the passage of so many years, except for the long hallway. Beautifully framed painting after painting hung there, all works of Neon Park, all the most unusual and brilliantly captivating 70s art.

There was a painting of Humphrey Bogart from the end of *Casablanca* standing on the tarmac wearing his trench coat and hat. Only in Neon Park's art, Bogart had an orange duckbill where his mouth and nose would be, and orange duck legs emerging from the trench coat and orange webbed feet.

"That one is called Bogart Duck," Neon said grinning.

"I love it, man," I said, flashing to the famous Bogart lines at the end of Casablanca: something about the beginning of a beautiful friendship, and something else about the problems of three little people not amounting to a hill of beans in this crazy world. Only in the movie Bogie pronounced it *whyold*.

"Yeah, man. I really love it."

"This next one is called Marilyn Duck," Neon said.

Indeed, there was an interpretation of Marilyn Monroe, in the pose from the famous 1955 calendar, naked on that red velvet bedspread, but with an orange duckbill, orange legs and webbed feet, and white tail feathers.

I looked at Buz. "I know you like this one."

He just stood there with the boa constrictor draped over his shoulders, smiling.

We saw other paintings of famous people adorned with duckbills, like Betty Grable in that over-the-shoulder pose from the famous World War II poster. And there were many album covers for a band called "Little Feat," and a really famous and slightly grotesque album cover that Neon did for Frank Zappa called "Weasels Ripped My Flesh."

We made our way through the hall of memorable album covers and duckbill art and were soon out back where the landscape was even more woodsy than in the front of the house, with trees, bushes and flowers, splendidly arranged like in a landscape design magazine.

Sitting on a bench was a couple, the woman who would be the star of the scene that Chick was filming, and an English dude who we soon found to be the boyfriend of the soon to be movie star. The woman was wearing a blue silk robe at this time.

There were some nice to meet yous and handshakes, and then Buz draped the snake around the woman's shoulders and suggested she get acquainted with it. She stood up and shifted the boa around on her shoulders. Then she looked at Buz.

"This is a lot heavier than I thought."

She was a tall (maybe five feet 8 inches) attractive blonde. (And a real blonde too, I couldn't help notice a few minutes later.)

I explained that this was the most docile snake we had ever had, and Buz showed her how to properly support it. Then I spouted the line I always say when someone touches or holds a snake for the first time, "You know, once you start liking snakes, you can never go back."

The young lady was shaking. She had never touched a snake before, let alone held a large heavy one like this. Neither had she taken her clothes off in public before or even acted in a film. Clearly there could be no filming until she became more relaxed.

I will not say from which direction, but suddenly a joint of marijuana appeared. Although this was for medicinal purposes to relax the movie star, not one of the rest of us wanted to be unsociable or unsupportive to the advancement of the upcoming film shoot, so we passed that joint around and became relaxed together, like a team. I may not be proud of that act right now, but this was the 70s, man, and each one of us clung to the axiom that the show must go on.

Soon thereafter, Chick lifted her large movie camera and asked the young lady, "Do you think you're ready to give it a try?"

She nodded, but when she stood up she said something about having a dry mouth. A glass of wine appeared and another obstacle was overcome.

Chick, the director, gave directions to the young lady. "Stand over here in front of this group of trees. Move around sensuously with the snake, dancing if you feel like it. Let's do a dry run first."

The young lady (I don't know if we ever learned her name, but I will call her Lola from this point on) stood up and handed the snake back to Buz while she undid the tie on her robe. She handed the robe to her boyfriend, the English dude, and Buz draped the boa constrictor around her shoulders again.

Lola was totally naked. To merely say that she was well developed, or had big knockers would be an understatement. Let's just say it would have been impossible for her to have fallen flat on her face. She shuffled to the area Chick had directed her to in front of some trees.

Chick pushed a button on a boom box and loud rock music filled the air from a group called *The Waves*. Lola was doing little more than slow dancing and staring at the snake. The boa, with iridescent shiny scales, was slowly moving through Lola's hands and around her body, constantly flicking its black, forked tongue. It seemed to me that the snake was the star of the scene at that point and Lola was too stiff to do much moving.

Meanwhile, Chick, who had said this would be a dry run, had in reality run the camera, I guess in case something she liked happened. But Lola's movement was not to Chick's liking, so she shut down the boom box and lowered the bulky camera from her shoulder. Lola was still shuffling back and forth and looking at the snake, which was moving around slowly, flicking its tongue and tucking its tail here and there.

Chick again gave directions to Lola. She asked her to try to move more to the music and to show on her face what she felt to have the snake moving along her body. But before Chick finished talking, Lola yelled out, "Ow! What's that?"

I looked and the big snake had dug the point of its tail right into Lola's thigh.

"Don't worry," I called out. "It's just the end of his tail. It's not a stinger or anything. It's harmless."

Lola's face was turning red. "Well could you get it off? I'm starting to freak out!"

I looked over at Buz, who was smiling. "She asked *you*, Richard."

"Fine, Buz, thanks."

I stepped over to Lola. She turned away from the group of spectators, and I pulled the snake away from her, gently manipulating it into a more comfortable position (for the snake as well as the naked lady). That's when I heard the shutter click on a still camera, and there was Buz taking a picture.

Richard Carefully Checks...

Richard carefully checks the welfare of his favorite Red-tail Boa during underground film shoot.

"Well," he said. "Chick said I could take a few stills."

"Yeah," I said, glaring at Buz and feeling my blood pressure go up. "Of her, not me, and definitely not of her and me together. Have you completely lost your mind?"

Buz started laughing. Then Lola started laughing. Then Marty-Neon started laughing. Even the English dude started laughing.

Chick immediately flipped on the boombox, raised her camera and in twenty minutes her scene was shot.

To me the art of Neon Park, whose real name was Martin Muller, belongs in the company of the era's greats like Peter Maxx, Leroy Neiman and Andy Warhol. In addition to creating the covers for almost every Little Feat Album, he also did album covers for Frank Zappa, The Beach Boys, David Bowie and Dr. John. His illustrations have graced the pages of The National Lampoon and Playboy. He passed away in 1993.

On the Saturday of that unforgettable film shoot, I knew almost nothing about Chick Strand. I was impressed with her friendliness and the way she went about her business. I didn't learn until much later that she taught film classes at Occidental College in Los Angeles for over thirty years.

I didn't completely understand what an underground film was. Okay, so this one had a naked woman cavorting around in the woods with a big snake. But that was just a scene. It wasn't dirty. It certainly wasn't pornographic. Actually it was beautiful, and for those of us who really like snakes, there was a spiritual quality to it. A few weeks after the shoot, Buz and I were given a gift of the outtake footage in sixteen millimeter film. We watched it a few times, but neither one of us has a clue where that reel is today.

A Pair of Boa Constrictors in Their Enclosure

The one on top is the movie star.

I discovered later that Chick Strand's body of work included over fifteen films, some of which are quite famous in independent film circles. In addition to being small independent films, perhaps the best description of underground films would be "experimental," although the earliest underground films often had "avant-garde," "psychedelic," or "counter-cultural" associations.

Although I never took the opportunity to see any of her films, I was saddened to learn of Chick Strand's death in 2009. A month after she passed, a tribute to her was hosted at the Egyptian Theater in Hollywood. Here are only a few of the phrases used to describe her work and herself: "integrity, honesty and selflessness," "creates lyrical representations," "questioning, admiring and honoring what she sees," "innovation of a new film language," "radically pioneering work," and "brings poeticism and the personal."

Chick Strand and Neon Park were soul mates. They were together for 30 years.

It was my lucky day when I was asked if I had a big snake and wanted to have an interesting Saturday afternoon.

*See "Plaster City" – _Snake Hunting on the Devil's Highway_

Strange Arizona City Law - Prescott

No one is permitted to ride their horse up the stairs of the county court house.

CHAPTER 9
Snakes in the Grass

When Buz unpacked his bag in Vietnam in 1966, he laid out a whole bunch of reptile books on his cot. Pretty soon all the guys were calling him *Snake Man* or *Doc*. Doc was fitting because he was originally supposed to serve as medical support, but they had too many personnel. So instead he was put in charge of the communications shack, where from time to time, he'd flip on an eight-track tape. Some of the favorites there were *Secret Agent Man* sung by Johnny Rivers; *The Letter* by the Box Tops, *Respect* by Aretha Franklin; *War* by Edwin Starr; *Crimson and Clover* by Tommy James and the Shondells; and *American Pie* by Don McLean, just to name a few.

Word quickly spread of Buz's interest and expertise. They saw that he had the books, and they observed him constantly picking up live snakes and lizards and moving them out of the way.

Lots of warnings were issued to the troops to be on the lookout for venomous snakes: cobras, bamboo vipers, and to be especially vigilant for a deadly mutha-effer known there as the "Three-Step Snake." Supposedly this monster was so deadly that, after being bitten by one you'd only be able to take three steps before dropping down dead in your tracks.

Soldiers would approach Buz all the time and ask, "What do I do if a *Three-Step Snake* bites me?"

Buz would look at them for a minute to let the tension build, and then he'd say, "Just don't take any steps."

Buz told me later that this phantom snake was likely one of the many pit vipers of the Trimeresurus species, like the Bamboo Viper. For the record there are many other snakes in Vietnam with more toxic venom than that one, including cobras and Kraits, but none can kill you within three steps.

One time in Danang, as part of the 5th Medical Battalion Fleet Marine Force, Buz was out in a graveyard with his unit. They had set up there because the enemy typically shied away from cemeteries. Suddenly he heard some chattering that wormed like a wave down the line until it reached him.

"There's a huge snake in one of the perimeter bunkers. Get Lunsford down here immediately!"

Buz went down into the bunker but didn't see anything. He started jumping up and down on a bunch of pallets trying to make some noise and vibrations. He still could not find the reported snake. Suddenly, he turned around and there it was, standing up so tall it almost looked him in the eyes, a very large cobra.

He knew he had to make a quick decision, and before he allowed himself time to think about it, he distracted the snake with one hand and grabbed it behind the head with the other. When he stepped out of the bunker with the large cobra he heard a combination of muffled gasps and cheers. He felt like a fisherman who had just landed a trophy fish in front of a crowd.

Buz didn't want the cobra to be killed so he carried it out to a rice paddy and released it. Only then did he permit himself the luxury of slumping his shoulders, taking several deep breaths and wiping the sweat off his face.

Another time (in 1967) he was on a patrol with about nine other guys. He was walking in the middle of the line and noticed all the soldiers in front of him stepping over a big log in the elephant grass. He looked down and saw a massive Reticulated Python ("Retic"), at least eighteen feet long. It had just eaten a pig, or other large animal, and was lying there all stretched out and perfectly still.

"Hey," he said. "That's not a log you guys are stepping over. It's a big python."

The head and tail were both hidden in the grass, but he could tell which way the snake was facing by the direction of the scales. Everyone stopped and watched as he went for the tail and dragged the entire snake out of the grass. The massive snake was lethargic and didn't put up much of a fuss.

Buz observed a couple guys taking pictures, but was surprised a short time later when one of the pictures of him and the snake appeared along with the story in *Stars and Stripes*.

A Snaky 18 Feet

Buz pulling an 18 foot long Reticulated Python out of the brush in Vietnam, as appeared in Stars and Stripes newspaper. Note bulge in snake from a recent large meal.

CHAPTER 10
Blythe's Giant Space Figures

"I'm just curious, Buz. Seems like every time we go snake hunting you tell me some crazy stories that you heard from this mysterious old wild-eyed rock shop miner dude."

Buz interrupted, "They're not wild stories. Every one of them is accurate."

We were sitting in The Space Age Coffee Shop in Gila Bend, Arizona, eating cheeseburgers and fries and drinking cup after cup of coffee. Buz had placed many used plastic creamers in a geometric formation on the table. My coffee was black. It was about 3:30 in the afternoon and we were hanging around until just before dark, when we would start looking for snakes south along Highway 85 to Why, and then proceed southeast

Sign on Best Western Space Age Lodge

Another location that adds to the alien spacecraft in the desert mystique is the Best Western Space Age Lodge. A regular stop for Buz and Richard is the Space Age Coffee Shop adjacent to the lodge

Photo courtesy of Steve Cotton

along the famous Ajo Road, Highway 86 to Tucson.

The Space Age looked altogether different back in the 70s than what it looks like today. The restaurant sign was shaped like a rocket ship. A similar sign designated the adjacent Best Western Space Age Lodge, and Sputniks sat on the roof. Inside the coffee shop the motif was more western than outer space. The restroom doors had signs in burnished rope formed into letters for *damas* and *caballeros*. In the parking lot up against the building were boulders with streaks of chrysocolla or turquoise running through them. I so wanted to take one of them home, but I couldn't have budged them if I tried.

The cast and crew of the 1973 movie, *The Man Who Loved Cat Dancing*, frequented The Space Age when the movie was being shot in that area. And the production of that film brought Gila Bend into controversy and international news when David Whiting, who happened to be lead actress, Sarah Miles', business manager and boyfriend was found dead in his motel room with "pills and bottles... scattered around his body."*

Sometime in the 1980s the lodge was refurbished. A large realistic flying saucer was placed on top of the lobby and the Sputnik satellites went bye-bye. In the late 90s the restaurant burned down, but was re-built with an updated space theme. Today there are still space age murals on the walls and futuristic metal tables, chairs and booths. The beautiful moon rock boulders are gone from the parking lot. You can still get a cheeseburger there, but they now specialize in Mexican cuisine.

Buz and I would often stop at the Space Age on our snake hunting trips, as it was the half-way point between Yuma and Tucson and the gateway to the Ajo Road. Sometimes we'd eat breakfast for lunch and drive up the road to the dam, where we'd take off our boots, roll up our pants and muck around in the shallow water looking for Sonoran Mud Turtles. Almost always we'd get more mud than turtles though.

We had gone there the year before and I had slipped and pretty well coated myself with mud. Buz had practically laughed out a lung, and I had gotten so mad that I grabbed a fist full of mud, almost like a dark gray snowball and hurled it at him. Naturally I missed, the throwing motion causing me to slip again which brought even more laughter.

But on this year's trip we devoured our burgers, drank more coffee than should be legally allowed, and completed our discussion on the veracity of the old rock shop miner dude's pronouncements.

There was silence for a few minutes. Darkness was not coming fast enough for me. Watching Buz play with the plastic creamers made me anxious. He was all the time changing the geometric patterns on the surface of the table. Finally, out of boredom I asked him, "Are you gonna tell me another story or wait 'til midnight when we're in the middle of freaking nowhere to try to scare me?"

"How many times do I have to tell you, Richard, they're not stories. But since you asked so nicely, and since we're sitting here in this restaurant with an outer space theme. . ."

"Wait," I interrupted. "You're not going to suggest that the space theme is no accident because, hmmm, let me guess, a flying saucer once crashed on this very site."

"No, that was way over near Tucson, and I've already told you about all of that.** I think the theme is because we were in a space race with Russia when they built this place and the motel next door."

Buz looked around anxiously for the waitress. "Hey miss, can we get some coffee over here?"

As he grabbed the last two creamers from the little bowl they were in, he began this new account:

"In 1932 a pilot was flying his small plane over the Colorado River when he noticed a group of huge figures drawn in the ground. The figures were of a cat-like animal, a circle with an attached spiral, and two large humans, the largest being over 170 feet from head to toe. The figures are so large, you probably could walk on them and not know they're there. They only make real sense from the air."

Here the waitress came with coffee refills. Buz had already taken the paper off the creamers and poured the white contents into his coffee. The waitress disappeared and Buz yelled, "Can we get some more cream over here!"

Buz saw that he had my attention and continued. "Nobody knows for sure where these drawings, which they discovered later are really carvings in the rocky surface, came from. But the archeologists think they're between 450 and 2000 years old. Some think that these designs were made by local Indians way back then, but there are those who wonder if the designs are not signs or tributes to ancient aliens."

"The old miner dude told you about this?"

"Yes he did."

"Have you seen them with your own eyes?"

"Absolutely."

"Okay, what does he think and what do you think?"

"Well, he believes that not only were these signs or signals to ancient aliens, but they were made by them as well. I only know that these are not like the petroglyphs you see on rocks or in caves. What other reason could there be for carving 170 feet figures in rocks where you can make out every finger, but you can't hardly notice them except from the air?"

"Well, where are these designs and when are we going there?"

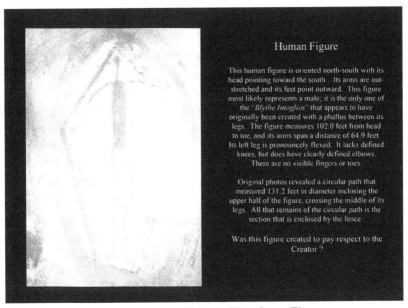

Human Figure

This human figure is oriented north-south with its head pointing toward the south. Its arms are outstretched and its feet point outward. This figure most likely represents a male; it is the only one of the "*Blythe Intaglios*" that appears to have originally been created with a phallus between its legs. The figure measures 102.0 feet from head to toe, and its arms span a distance of 64.9 feet. Its left leg is pronouncedly flexed. It lacks defined knees, but does have clearly defined elbows. There are no visible fingers or toes.

Original photos revealed a circular path that measured 131.2 feet in diameter inclosing the upper half of the figure, crossing the middle of its legs. All that remains of the circular path is the section that is enclosed by the fence.

Was this figure created to pay respect to the Creator?

Interpretive Sign at Blythe's Giant Space Figures

This human figure is oriented north-south with its head pointing toward the south. The arms are outstretched and its feet point outward. The figure most likely represents a male. It is the only one of the "Blythe Intaglios" that appears to have originally been created with a phallus between its legs. The figure measures 162 feet, head to toe, and its arms span a distance of 64.9 feet. Its left leg is prominently flexed. It lacks defined knees, but does have clearly defined elbows. There are no visible fingers or toes.

Original photos revealed a circular path that measured 131.2 feet in diameter enclosing the upper half of the figure, crossing the middle of its legs. All that remains of the circular path is the section that is enclosed by the fence.

Was this figure created to pay respect to the Creator?

"They're about fifteen miles north of Blythe just off Highway 95. We can probably go from Tucson to Phoenix and then on to Blythe instead of going home through Yuma. I'll show them to you. You want to know what I think?"

"I'm dying to know."

"I'm not so sure the drawings they say represent humans are really humans. The one they say is feline I'm pretty sure is some alien creature. And the circle spiral design has to be a sign or signal to entities from another world."

Buz paused here and finished his coffee. "Now I'm going to the caballeros' room. See if you can get that waitress to bring more coffee. One more cup and we're off to get snakes."

*Trivia IMDb, *The Man Who Loved Cat Dancing* (1973) Did You Know?

**See *Weather Factors*, <u>Snake Hunting on the Devil's Highway</u>

Giant Space Figure From the Ground

Buz: Some think that these designs were made by local Indians way back then, but there are those who wonder if the designs are not signs or tributes to ancient aliens.

Marshall Trimble
(Arizona's Official Historian)
Quirky Fact

Arizona's first mobile home was housed by William Hardy whom Hardyville on the Colorado was named for. He drug a steamboat's cabin ashore and took up residence in it.

Movie Quote – Snakes on a Plane – 2006

Neville Flynn: Enough is enough! I've had it with these motherfucking snakes on this motherfucking plane?

Movie Quote – Snakes on a Plane – 2006
(Alternative Line for Television)

Neville Flynn: Enough is enough! I've had it with these monkey-fighting snakes on this Monday-to-Friday plane!

CHAPTER 11
Muffin for Dinner

Muffin was a beige colored miniature poodle dog that came into the family as a reward for finding and caring for Muffin's mother, who was lost in the neighborhood. We had kept Muffin's mother with us for a couple weeks and the entire family became very attached to that sweet dog. None of us wanted to give her back to her owners, but when they responded to our "Found Dog" ad we knew we had no choice. Muffin's mother's family was so happy to get her back, they promised us pick of her next litter. That's how we got Muffin.

For a poodle, Muffin was adventurous, but not exactly courageous. One night we heard him barking at the patio door. It was not Muffin's usual behavior to get that excited, so we wondered what was up in the backyard. We switched on the patio lights and slid open the glass door. Muffin tore out and ran right up to a huge opossum, barking viciously all the way. The opossum hissed loudly and barred its teeth. That's when Muffin cried out a series of four poodle yelps in deep fear, wheeled around, an expression of embarrassment and disappointment on his beige poodle face, lowered both his head and tail, and hurriedly four-legged it into the house, where he slumped down in a corner, clearly wanting to be alone.

Muffin was a great dog, but he was crazy about, and completely devoted to Irisse, and would only come to me if she was busy or not around. Nonetheless, I'd sometimes take him for a walk and see other guys walking their pit bulls, Rots and Dobies. Inevitably their dogs were well trained and would obey commands with pride and efficiency. Muffin, on the other hand, was all over the place. He wouldn't listen for anything, especially when anyone else was around.

Years before professional snake-keeping rack systems and custom cages were widely available, I hired a carpenter, gave him a set of crude drawings and had him build a unit of 27 cages of various sizes. They were backlit behind thick Plexiglas, had heat bulbs in each cage and individual locking glass doors. The entire unit stood just shy of wall to wall in the "snake pit," as my combination office, library and reptile room came to be called. People who saw the unit were usually impressed by its beauty as a piece of furniture and functionality as an apartment complex for snakes.

At about waist level on the left side of the unit was a large cage for Georgia, the same Burmese python which had paid Buz a visit in the hospital several years before. Now Georgia was a teenager, at least nine feet long and quite hefty.

It was feeding time for Georgia and I was listening to a Ricky Nelson tape while opening her cage door and tossing in a live rat.

Georgia quickly killed the rat, swallowed it, and I tossed in another. I repeated that process five times. Then I had another live rat in my hand and I was getting ready to offer it to Georgia when Muffin pushed through the slightly open door and walked over to me, panting for attention. *Oh no*, I said to myself. *This is not good.* But my hand was full of rat and the cage door was slightly ajar.

What happened next occurred extremely quickly, but at the time, it seemed like I was in a slow motion time warp. I knew I had three things to do: get rid of the rat, close the cage door and get Muffin the hell out of the room. But panic and fear slowed me down to the point of not being able to do *anything*. I showed Georgia the rat and started to toss it in her cage. Unlike me, Georgia was in a triple-fast-forward-feeding-frenzy. She saw the rat, smelled the rat, looked like she was ready to meet, greet and eat the rat. However, as soon as I gently swung it (by its tail) into the cage, Georgia's head turned ever so slightly and she noticed Muffin. She instantly must have realized that repeating the rat thing over and over was getting tedious. *Why not*, she must have thought, *eat one large meal and get it over with?*

I sensed what was about to happen and yelled at Muffin to get out of the way as I moved toward him, the rat still in my hand. But it was too late. Georgia gave a giant lunge out the partially open door and grabbed Muffin, who yelped out in surprise and fear.

Oh my God, I thought, as I finally tossed the rat into the cage. Fortunately Georgia only had Muffin by the fur and not the skin, and because Georgia was only partially out of her cage, she was unable to wrap Muffin in even one coil. But she had Muffin nonetheless and I could see her trying to get in position to throw coils around the yelping poodle. Still in slow motion, I came at Georgia and gave her a hard, open-handed blow to her neck, using both hands, my fingers spread wide. Instantly Georgia released Muffin, who, yelping continuously, raced out of the room.

That was great news for Muffin but bad news for me, because Georgia turned to me and struck hard, sinking her large curved python teeth into my right hand. Taken by surprise I yanked my hand backwards and Georgia let go. Somehow I got her back in her cage where she turned her attention to the rat.

I slumped down on the floor and thought about how I almost had to tell Irisse that the dog she loved so much had just become python chow. No, I probably would have had to play dumb and suggest that maybe Muffin just got out of the house somehow and ran away. Or maybe *I* would have just run away.

Muffin

Muffin survived becoming python chow and poses here in his best holiday outfit.

Then I noticed my hand, which felt like it had just been hit with a shotgun blast, and was now bleeding like a mofu. (Yanking my hand back was a stupid thing to do, as it caused the wounds to rip.) Then Irisse ran into the room screaming something that I didn't understand for a few seconds in my fuzzy-slowed-down world. And when she noticed my hand she really started in on me.

I finally snapped out of it and heard everything she was then screaming. I'm crazy for keeping these damn snakes, and why is Muffin so freaked out hiding in a corner and still yelping, and I'm bleeding on her clean tile floor and...

Well, I told her that I know I'm crazy by most people's standards, and Muffin was freaked out because I forgot to close the snake pit door all the way when I was feeding Georgia, so it was my own damn fault that Georgia grabbed Muffin when Muffin happened to walk in at the exact instant I was tossing Georgia a rat, but Georgia barely grabbed Muffin by the fur and Muffin only got the crap scared out of him and didn't really get hurt because I got Georgia to let go, only when Georgia let go of Muffin she swung around and bit the crap out of me and that's why I'm bleeding like a mofu on your clean tile floor!

Movie Quote
Snakes on a Plane – 2006

Neville Flynn: "Everybody listen!
We have to put a barrier between us and the
snakes!"

CHAPTER 12
Alien Snake Abduction

It was a warm, humid evening in August. Dark monsoon clouds had rolled over the Chiricahua Mountains earlier and drenched the entire Sulphur Springs Valley. That wet desert smell was once again in the air, fragrant with the promise of good things to come. We were on Highway 186, going east toward the mountains out of Willcox, Arizona. I was driving and Buz was shotgun.

I dialed through the radio and stopped when I heard Paul Simon singing something about the days of miracle and wonder.

About 8:15 Buz said in a calm voice, "Aren't you going to stop for that snake back there?" His tone indicated that he was sure it was a snake and not a broken fan belt, piece of rope, fast food wrapper or donkey turd, and I felt mildly humiliated. How could I have missed the first one?

The road was deserted so there was no problem making a u-turn and slowly coasting onto the shoulder, where the obviously (by its size and shape) non-venomous snake was basking. Buz quickly got out and grabbed the twenty-six inch long-nosed snake and brought it into the car.

"You saw that on the shoulder," I said. "Nice call."

He didn't rub it in, which was unusual for Buz. Instead I noticed that he was spending considerable time examining this rather common snake with the flashlight.

"You know," he said, "there's a guy I know in Florida who wants a real nice male. He wants to trade a Florida Kingsnake for a good specimen, and this one looks prime." It was a healthy example with nice red and black saddles on a clean buff background.

I stretched out the tail. "I'm not positive, but this looks like a female."

"I think you're right," Buz said, "but I'd like to make sure. Let's use a small probe. If it's a female, we'll set her off the road."

So I held the flashlight and the pretty snake while Buz did the probing. Then he turned to me and said, "You're right, it's a female. Let's release her and see what else is out tonight." We walked about twenty yards through the sand and gently set the snake down. It crawled quickly away.

We drove with the windows down. A few minutes later moths began splattering on the windshield, attracted to the headlights. I rolled my window up.

Buz started laughing. "What's the matter, Richard?"

I belched. "You know very well I don't want any moths or anything else that flies around in the night getting in the car. I'm sorry for them when they hit the car, but I don't need them fluttering around my face."

Instantly my thoughts turned to the time I was driving on the busy interstate in the middle of the night. Buz was asleep. I felt something definitely crawling up my leg inside my jeans. But wait, that comes later in this book. See *Don't Bug Me, Man.*

I was brought back to the present when a nighthawk swooped gracefully down and picked off a moth in the headlight beams.

Suddenly I started thinking about that female long-nosed snake and UFOs.

"Can you imagine," I said, "that snake going back and telling all the other snakes about what she just experienced? It's like all these guys who say they were abducted by aliens.

"Hey Alice, you won't believe what I went through last night. This huge spaceship landed on the road not far from my rock. I was just lounging outside when I saw these bright lights moving real slowly in my direction. I tried my best to lie still and blend into the sand, and I thought they didn't see me, but they turned around and stopped right in front of me. I tell you, Alice, it was horrible. This giant creature grabbed me and took me into the spaceship where two unspeakably hideous alien monsters shined an intensely bright light and examined me in the most obscene manner. It was so humiliating. They spoke a strange disgusting language. It gives me the chills to even think about

it. I don't remember if I was there for a few minutes or a few hours. I thought they were taking me to another planet, but I was surprised when they carried me off the road and just let me go. Then I saw the spaceship take off in a cloud of dust."

It was at that point in relating this that I paused and looked at Buz. "So then Alice says, *Look Susan, we've been gossip buddies for a long time. You know I believe you, but if I were you, I wouldn't go around telling all the other noses about all this.*"

I paused and looked at Buz. "I figure *noses* is what a long-nosed snake might call others of its kind."

Long-nosed Snake

"Hey Alice, you won't believe what I went through last night!"

Then we stopped to grab a juvenile Sonoran Gopher Snake, which hissed in protest, but did not show teeth. After we measured and released the little hisser, and recorded it in our log, I wanted to continue our abduction discussion. "What do you think?" I asked. "Isn't that just the way it happens?"

Buz chuckled softly and said, "Yes, Richard, that's the way it happens, but you're forgetting that snakes have no ears, so they couldn't hear our disgusting language, and they don't really speak other than the variations of the sound that gopher snake just made. Besides, why would Susan speak English, even if she could speak? And how would she know about other planets. I mean, did she go to snake school?"

"Damn it all, Buz. Don't you have any imagination?"

"Sure I do. I'm just giving you a hard time. And while I'm at it, I am of the opinion that aside from parrots, mynah birds and a few others, why would any animals want to talk to humans anyway? Those that we don't eat, humans are often mean or abusive to, even our canine and feline pets. Homo sapiens can be such assholes to animals."

"Right you are, Buz. I can't even look at the many abused animals on internet sites or on t.v. commercials asking for money."

There wasn't much out that night, and I had a few hours to obsess about the whole spaceship thing. The road was deserted at about 1:00 a.m. and we pulled over for a pee break. Finally I said, "Well, if Susan did go to school I bet she was in kinder-*garter*, ha ha ha."

Buz looked my way and smiled. Then he said, "I think you were the one abducted by hideous aliens, and they zapped you with an obsession ray and now you won't stop talking nonsense."

"Well it could've happened. How do you know that snakes can't communicate with each other by non-verbal means, or..."

"See what I mean? You've been zapped and you won't stop."

That's when I finally shut up.

Galaxy Peace Patrol

"Do you know what's weird, Buz?"

"No, What's weird, Richard?"

"These guys are parked in California...

"...but their vehicle is registered in Nevada!"

Alien Fresh Jerky, Baker, California

CHAPTER 13
Drop That Snake!

When my middle son, Mickey, turned thirteen, I took him on a father-son vacation to Orlando, Florida to do the whole Disney World/Epcot Center thing. By day we would experience the theme parks and by night relax in a nice motel room, watching t.v., listening to music, talking about the day's activities and our plans for the following day. You know, just hanging out.

Only on this particular night, Mickey wanted to go play video games in the motel's arcade. Okay, so he was thirteen years old, I thought maybe he'd meet some kids. So I sat around listening to music on the radio. At the time I heard the doorknob jiggling, the melodic sounds of *Karma Chameleon* by Culture Club were blasting through the room.

As the motel room door opened I was thinking how coincidental it was to hear the word *chameleon* in a song, as I had observed a couple of anoles basking on branches in the motel's courtyard earlier that day, even though it was late in September. I explained to Mickey that we don't have these lizards in California, but that when I was a kid, I kept several as pets and marveled at how they changed colors like the true chameleons.

But now the music was blasting and Mickey was standing in the open doorway with a huge smile on his face. "Look what I caught," he said.

Only then did my eyes lower to the squirming snake in his hand, yellow, red and black colors swirling too quickly for my tired Disney World eyes to focus on.

"Drop the snake!" I said loudly enough to be heard over the chameleon music, but not too loud where Mickey might panic unnecessarily.

Mickey froze. "What?" he said too softly to be heard over the music.

I still could not focus on the pattern on the small squirming snake in his hand. I was convinced that Mickey would let go of the snake this time, so I didn't move. "Just put it on the floor for a minute, Mickey! Please."

Now Mickey was totally frozen in place and I began to freak out. I realized that it was probably a "red to black –venom lack" species, but I was a California native and had never seen an Eastern Coral Snake close up and personal before. The snake was still thrashing about in Mickey's hand which was clamped onto its actively moving body like a vice. Mickey noticed the worry on my face and he suddenly appeared concerned too, but he was still frozen in place.

"Drop the goddamn snake!!!" I yelled and made a lunge for it, somehow knocking it out of his solid grip.

The snake began to crawl away slowly, not crazy like the way it had been thrashing, and I could clearly make out the pattern. Definitely not venomous, I quickly assessed, and probably a Scarlet Kingsnake.

"Okay," I told Mickey, who appeared confused and a bit angry, "Now you can pick it up."

He said, "Why did you do that?"

I said, "Because I couldn't see the pattern and it could have been a Coral Snake. They have cobra venom. I love you and wanted to make sure you were safe. Now go catch that kingsnake again, it's a great specimen!"

Mickey picked up the snake before it got away and handed it to me. He told me how he had seen it crawling in the motel's courtyard and just picked it up. I told him that I've never caught one of these, that it's a fantastically beautiful snake, although definitely on the nervous side and that he did real well to have caught that beauty. I explained that this little snake could have caused a real panic had some of the other tourists encountered it before he did.

Then he looked at me and said, "Can I go back to the arcade?"

As I slipped him a five dollar bill, I said, "Sure, knock yourself out. Just don't bring back any water moccasins."

When Mickey and I walked through the big alligator's mouth at the entrance to the Gatorland Zoo* in Orlando, I had a strange sense of deja vu. While pondering this odd feeling, a movie called *Lake Placid*** came to mind. A giant 30-foot long crocodile goes on a man-eating rampage in a fictitious Maine lake, leaving blood and gore everywhere. Indeed the proportion of the open alligator's mouth at Gatorland's entrance seemed about right for a monster movie croc that was able to devour whole an entire cow.

Richard Holding a Young Alligator

At Gatorland Zoo

Shaking those images from my consciousness, I followed Mickey who seemed to be motivating on his own course through the park. I clicked pictures of him in the open mouth of a large fake gator on a poolside deck, of captive gators in the water, out sunning themselves on land and in a natural-like swamp. I paid five bucks to have a Polaroid taken with me holding a young live gator, unfortunately with its mouth tied shut, I suppose for liability purposes. We watched a feeding show and took a short hike on a raised wooden pathway through the Gatorland swamp. Then we took some pictures of live Eastern

Mickey, Please...

"Mickey, please get out of that fake alligator's mouth and let's go check out the venomous snake pit." At Gatorland Zoo, Orlando, Florida

Diamondback Rattlesnakes, cottonmouth water moccasins and copperheads in the venomous snake pit.

At one point Mickey asked me what the difference was between alligators and crocodiles. I explained that alligators have wide heads, shaped like a "U," while crocodiles have narrower, more pointed heads, shaped more like a "V." I told him that gators prefer to live in fresh water environments while crocs often thrive in saltwater habitats. I mentioned that you can see a protruding fourth tooth on the lower jaw of a crocodile when it has its mouth closed. I asked Mickey to look at some of the alligators around us to see if he could see any such protruding fourth tooth. He could not.

"Where do alligators go when it gets cold, like in hurricanes and stuff?" Mickey wanted to know.

"Well," I told him, "All reptiles slow down when it gets cold. Snakes and lizards will hibernate. Alligators will dig a hole in a riverbank or at the edge a pond and just go dormant. They can dig in the mud or just chill and wait for a sunny day. When it does get warm they will come out, but they won't eat unless it gets hot enough."

Then I explained how alligators were on the endangered species list for many years because their hides were used as exotic leather for boots, shoes, belts, ladies' handbags and for briefcases and luggage. And I told him that even while they were protected, a large amount of poaching was going on. But thanks to law enforcement and habitat protection, they were removed from the endangered species list around 1987. Today they thrive.

"Alligators are cool, dad," Mickey said. "Can we get one?"

"No," I replied. "We couldn't give it proper space or proper care. Only thing we can do is appreciate them in the wild or in places like this."

"Okay," he said. "Then can I get a soda?"

"Sure."

*Gatorland Zoo, a must-see attraction in Orlando, Florida. Address: 14501 S Orange Blossom Trail, Orlando, FL 32837

**Lake Placid, Fox 2000 Pictures and Stan Winston Studios, 1999, written and produced by David E. Kelley, directed by Steve Miner

Juvenile Alligators on Deck

At Gatorland Zoo

Adult Alligators in the Water

At Gatorland Zoo

CHAPTER 14
Beware of Desert Gastronomy

When you spend time in the desert hunting for snakes, from time to time you need to stop and have something to eat. Over the years I've learned quite a bit about how to do this, and more importantly, how not to do it.

Of course, you may bring your own food. This is a good idea only if you discuss the menu in advance and everyone in your snake hunting party agrees. Don't tell your buddy, "Okay, I'll bring the water, soda, beer and bread; you bring stuff to make sandwiches and some chips and snacks." That's what I did the first time I went to Arizona with Buz. He brought nothing but Spam, and that really freaked me out.

I believe there's a clause in Murphy's Law that stipulates if you *do* enjoy Spam, your buddy will bring something like wheat germ, or that horrible stuff that everyone spreads on toast in Australia. So I suggest that *you* bring the substance to make sandwiches and have your buddy bring the other provisions, because in the long run what difference does it really make if you end up drinking Coke or Pepsi, Budweiser or Coors?

Better yet, consider that you're on vacation, and if you've got the coin, you take your meals in cafés, truck stops, drive-ins and desert dives. But if you *do* eat out, you might want to take note of the following recommendations that Buz and I compiled over the years:

If you're driving through the desert and you see a sign on a building that says "EAT," don't. Same advise if the sign says "FOOD." It's not only the cuisine in these places, or the ambiance, or the abundance of (non-reptilian) desert creatures lurking about. By nature, these dives are isolated in the middle of nowheresville, and are the perfect place for zombies to attack, as evidenced in the movie called *Legion.** "...a dramatically muddled but surprisingly involving Bible-themed fantasy thriller that imagines Armageddon in a dingy roadside diner...”** I don't know about you, but when I'm on a snake hunting trip, I don't care to watch people turn into ghouls and crawl up the walls and dangle upside down on the ceiling as I'm trying to enjoy my cheeseburger. And I really don't need to see the owner of the place take the handy shotgun he's got stored under the counter (I guess for just such a purpose) and blow the zombie away, splattering blood like much more ketchup on my fries. No sir, I'd rather eat my greasy burger in peace and quiet.

Another thing to keep in mind is if the menu lists “plate-size” pancakes, do not make the mistake of questioning the size of the plate, like Buz's nephew once did in Why, Arizona, at the gateway to the Organ Pipe Cactus National Monument.

This is how that scene played out:

Raymond to Old Crotchety Lady Waitress: It says here *plate-size* pancakes. Well, what size are the plates?

Old Crotchety Lady Waitress: (Staring at Raymond in disbelief for at least fifteen seconds, then turning away and yelling loudly): Blanche, this guy wants to know what size are the plates! Haw, haw, haw!!!

Naturally, all the other customers (2 of them, I believe), along with Blanche and our Old Crotchety Lady Waitress couldn't stop laughing and slapping their knees the entire time we were there. Buz laughed too; but I slumped down in the booth, embarrassed about half to death, and Raymond walked out and sat in the ridiculously hot car pouting.

The curious reader might wonder where Why, Arizona, is and how it got its name. Or in other words, where and why is Why? Why is ten miles south of Ajo and thirty miles north of Mexico. It is also just north of the Organ Pipe Cactus National Monument.

Herpetologically speaking, on the north-south road (Highway 85) between Why and the Mexican border you can find Gila Monsters and the chocolate striped Mexican Rosy Boa. However it is illegal to collect anywhere within the boundaries of the National Monument, and that comprises a good deal of that road. Not only that, but the entire area is heavily patrolled by all manner of law enforcement, being so close to Mexico. My advice to anyone compelled to collect in that area: Be very, very careful, and do not stop to pick up any 2-legged creatures under any circumstances. These days it is beyond dangerous, unfortunately.

Concerning dining in the desert, another thing to consider is if you have advance notice that the area in which you're intending to hunt snakes has only one café and one hamburger stand, and you've been told by other collectors, who really should know what they're talking about, that a wolfman owns the café and a cyclops runs the hamburger stand, well, you'll have to make up your own mind if you really want to risk dining in either of those establishments. You might ponder that the aforementioned scenario is ridiculous, unrealistic, would never happen in real life and what kind of drugs

Why, Arizona

Perhaps not as amusing as Bumble Bee or Total Wreck in Arizona (see the many other strange and awesome Arizona place names throughout this book); Bummerville or Rough and Ready in California: Elephant Butte or Truth or Consequences in New Mexico; or Hurricane or Nipple in Utah, but Why, Arizona still has its own story. Originally, Arizona state routes 85 and 86 intersected in a Y junction right where the town was founded. When it came time to name the town, it was discovered that in Arizona, it was required that a town's name must be at least three letters long. So instead of the town being named Y, the founding fathers decided on Why. The entire exercise was for naught, however, because some time later, the Department of Transportation changed the intersection of the two highways for safety reasons into a traditional T. Apparently it was too late to change the town's name to Tee.

was I on to even suggest such a thing? While I don't blame you at all, I must insist that, not only is this situation possible out in the dark of the desert, but that it actually did happen to me.

In the mid 1960s the population of Borrego Springs was only about 300***. Completely surrounded by California's largest state park, the Anza-Borrego, Borrego Springs was a prime destination for campers, hikers, naturalists, amateur astronomers and herp enthusiasts. I started going there in 1965, because all the older herp guys told me that that's where the snakes were. But some of them also warned that a wolfman owned the café and a cyclops ran the A & W Root Beer stand. When I told my parents about this, (I was only sixteen years-old at the time), my mother said that I definitely should not go, but my dad suggested that these older guys were just messing with the *kid* and I shouldn't let them stop me.

So I went. When I got there I immediately started asking questions in the gas station, the store and everywhere else. I discovered that the owner of the café was none other than Lon Chaney, Jr., who played the original wolfman in the movies. That was a relief, because I thought movie wolfmen were cool. Even so, I still avoided that place. Turns out that the so-called cyclops was merely an old desert dude with a patch over one eye.

While most folks going snake hunting today probably wouldn't give a thought about wolfmen and other assorted monsters in the desert, some of us geezers who grew up in the 1950s were heavily influenced by the abundant science fiction and horror flicks, which, along with westerns, dominated the silver screen in that decade. To this day, I can't drive through the desert without thinking about the huge mutant (usually because of atom bomb testing) creatures, like car-size killer ants from the movie, *Them*, or haystack-sized tarantulas from *Tarantula*. Each of these, and so many others were filmed with the signature Joshua Trees of the Mojave Desert all around. Couple the now lodged-in-my-brain desert monster images with my natural paranoia and you have an explanation of my constant habit of looking around for more than just snakes.

If you can get beyond the movie monsters, you're still going to need to eat. Here's something else to consider. Maybe you'll run into some friendly people who will invite you to their campsite for dinner. Buz and I did that one time, after meeting some people hiking trails in the Chiricahua Mountains. They were a poor religious group staying in a beautiful clearing in the pines called Methodist Camp. They were serving some kind of desert stew, but it was full of tiny white worms, and the bread that went along with it was moldy. Believe me, that was a "want to get away?" moment.

One final bit of advice. As you're driving across the desert, you'll see billboards now and then advertising delicious-looking food and beverage. You might see such a billboard around the Palm Springs/ Indio area along California Highway 10 advertising a Date Shake. No matter how appealing the notion of savoring that concoction seems, you must use all your powers of self-control and zoom past the nut stand that offers it for sale. The inquisitive reader may ask why. To say that the blending of ice cream, milk and date substance produces near lethal farts is a vast understatement. Enough said.

On one trip Buz and I stopped at that nut stand. We pulled into the nearly empty dirt parking lot in an oasis of palm trees. We were there for refreshment that did not include date shakes. We sat at the counter and were surprised to see that we were the only patrons there. Almost immediately the old lady waitress flopped a dark rubber rattlesnake on the counter and said, "Look what I found out back." She looked at each of us in turn, hoping for us to jump or scream, but she had no idea who she was dealing with.

Buz held up one finger and got up, which I interpreted to mean that he would be right back with something a bit more exciting than that limp length of rubber. I smiled in anticipation, trying hard not to laugh.

A couple minutes later Buz flopped a small, lightly colored, but thick-bodied snake on the counter, which immediately coiled and rattled intently.

"My God," the old lady waitress cried out. "Is it alive?"

"It's more alive than yours," Buz replied. He then pinned the sidewinder with a spoon, picked it up and we walked out of the nut shack, feeling much more satisfied than a greasy meal could have provided.

*Legion, 2010, Bold Films, Directed by Scott Stewart

**Quote from Joe Leydon, *Variety*

***Population of Borrego Springs in the 2010 census was 3229

Marshall Trimble
(Arizona's Official Historian) Quirky Fact:

For many years the town of Nothing, population four, had a saloon called the "Ain't Much." It burned in 1988 and for a while there was nothing in Nothing.

CHAPTER 15
Cooking in Baker

In Las Vegas you can drive along the "Strip" (Las Vegas Blvd.) and see the Statue of Liberty, the Eiffel Tower and the Great Pyramid. You can ride a roller coaster around the top of the Empire State Building, watch pirates trying to sink each others' ships and observe jets of dancing water shoot high in the air, choreographed to Elvis Presley's *Viva Las Vegas.* In Las Vegas you can ride a formula one race car, shoot an Uzi, get married in a drive through, or get a lap dance in a titty bar. Of course you can pull the handle, double-down, let it ride, take the odds or put it all on red. And remember, what happens in Las Vegas stays in Las Vegas.

World's Tallest Thermometer

Baker, California

Anyone who has driven to Las Vegas from Southern California knows that the tiny town of Baker (population 600) is about halfway between Barstow and Las Vegas on Interstate 15. Baker was established over a century ago to serve the needs of desert travelers. Now, you can stop there for a burger, some tacos or Greek food. You can fill your gas tank with expensive gas while gazing at the "World's Largest Thermometer." You can sample some out of this world jerky. Best of all, you can observe a variety of extra-terrestrials.

Baker also promotes itself as "The Gateway to Death Valley." If you are so inclined you can get to Death Valley by driving 117 miles from there on SR127. Can you hunt for snakes on that road? Yes, but I don't recommend it. Buz and I tried it a couple times with no good results. There's much more traffic than you would expect in both directions. There is also a loud annoying humming noise from farm equipment that can be heard for many miles, even with the windows up. And there can be an eerie haze or fog around all the agriculture. I don't know exactly what that is, but it doesn't seem natural. There are better places to hunt for snakes in the area.

Baker, California

Gateway to Death Valley

Baker has recorded some of the hottest temperatures in the country. In 1951 it reached 124 degrees in both July and August. I don't know about you, but I've driven through the desert with temps like that, and it's no fun when you have to stop for gas. As bad as it is at 124, neighboring Death Valley reached 134 degrees in 1913.

Baker also boasts that it has "The World's Tallest Thermometer." Travelers along Interstate 15 can note how hot it is as they drive by the 134 feet high tower. Since its erection in 1991, there were periods of time that this landmark recorded questionable high temperatures, which according to rumors were designed to bring media attention to the town. In recent years there were also periods where the thermometer was broken and recorded no temperatures at all.

Now if you're in the mood for food when you come to Baker, there are several fast food facilities. And there is one place that was visited by Guy Fieri and received good reviews on the Food Network's *Diners, Drive-ins and Dives*. That is the Mad Greek. So if you'd like to have a gyro, or some souvlaki, or maybe a taste of tzatziki when you're in the middle of freaking nowhere, and you'd like to consume it while

The Mad Greek Cafe

Baker, California

surrounded by white columns and other Greek glitz, man, this is your place.

Personally, I don't know what any of that stuff is, and the names scare me. And by the way, how can I be sure that there isn't some crazy Greek guy in there running around in a chef's hat and swinging a meat cleaver? I mean, I don't think that would happen, but it's on their sign (Mad Greek), and I don't know if *mad* in this case means pissed off or full-blown crazy. I figure that even with only a small possibility that there could be something to it, why take a chance?

Alien Fresh Jerky

Baker, California

When Buz and I are snake hunting, it's got to be fast food, because we have no time to waste, and any of the fast food eateries in Baker will do. Buz and I agree that it's about the snakes, not the food.

There is a place in Baker that I never fail to visit when going through the town. It's called *Alien Fresh Jerky*. This is an alien themed jerky joint, which, besides some rather unique jerky varieties, also sells nuts, candy, stuffed olives, a huge variety of sauces (mainly scorching hot) and unique beverages (like a red concoction called True Blood). They offer free samples of jerkies they call Premium Road Kill Original, Premium Colon Cleanser Hot, and Premium Weed Killer. They also have the more traditional ones. Most of these sell for $8.00 per 4 oz. package, or 3 for $20.00.

They also feature a huge selection of hot sauces, most of which, by glancing at their labels, appear to be nuclear hot. Two that caught my eye are "Ass in Space" and "The Hottest F***in Sauce," the latter of which proclaims to be "hot as f***."

As you might imagine, I don't go to Alien Fresh Jerky for their Road Kill or to purchase a bottle of sauce that can cause spontaneous combustion or one that could liquefy my sinuses. Unfortunately Buz and I encounter enough real road kill while snake hunting to detour us from even the free samples. No sir. Same with the Colon Cleanser and the Weed Killer jerkies. I have no doubt that they are original, and I don't care that they are premium. Buz and I don't go to Alien Fresh Jerky for the jerky. We go for the aliens.

Select Few Examples...

Select few examples of the hottest hot sauces available at Alien Fresh Jerky, Baker, California

Alien Fresh Jerky

Baker, California

The reason I am so infatuated by the concept of space aliens is that my consciousness is saturated with them from years of reading and viewing realistic sci fi literature. Add to the mix hundreds of urban legends and tall tales of UFO crashes, landings and human abductions, provided to me courtesy of Buz during quiet moments on nearly all of our snake hunts through the years.

Understanding that these stories were usually designed to freak me out does not diminish the effect. And living only a couple hours from Area 51, probably the most secretive and controversial government facility you could ever find in the desert, does not help either.

One time on a snake hunt, after listening to Buz say something about ancient aliens and their interference and influence in Mayan civilization, I asked him why he was so fascinated with space aliens and UFOs.

Buz was driving at the time. He shifted in his seat and I saw a smile evolve on his face. "Why am I fascinated with life itself?" he replied. "It's because they exist. Think about it. How could we on Earth have the audacity to think that with millions and millions of planets, we're on the only one that has intelligent life? It's about time we stop allowing ourselves to be influenced by a bunch of scientists who sit behind their desks and make the rules. The world should keep more of an open mind."

It was then that I reminded Buz that when we first met he told me that he was from outer space, and I asked him, "Do you still maintain that you are an alien from outer space?"

"You don't believe that?" He replied. "How could somebody so stupid have an answer for everything?"

Nobody that knows Buz thinks he's stupid. But he does have an answer for everything, and he is unique enough to make you wonder, what in the universe is he?

*Although broken for many months, as of October, 2014, the world's tallest thermometer is working again.

One of These Guys...

One of these guys says he's from outer space but enjoys snake hunting on Earth. The other guards a display at Alien Fresh Jerky, Baker, California.

Friendly Alien with Big Eyes...

Friendly alien with big eyes, and pissed-off alien with long fingers at Alien Fresh Jerky, Baker, California

CHAPTER 16
The Strangest Road

Just six miles west of Baker you can hunt for snakes on the strangest road, both in name and history in all the vastness of the Mojave Desert. The road is called Zzyzx.

Zzyzx Road is long and narrow like a snake, and it winds about four and a half miles around some low hills, ending at a surprisingly beautiful picnic area and research station. A self-proclaimed minister and radio commentator named Curtis H. Springer went there with his wife in 1944. Deciding that the area around a natural spring and prehistoric quarry would be perfect for a health resort, he immediately filed mining claims on over 12,000 acres of public land. He planted some palm trees, built his health facility and named it Zzyzx, promoting it as "the last word in health and the last word in the English language."

Springer bottled some of the spring water and sold it to hot, thirsty desert travelers. He made a fortune selling medical products that he claimed would cure everything from the common cold to cancer. One of his products he said would cure baldness. The healing powers of the natural hot springs which old folks traveled hundreds of miles to use were not natural as advertised, but were heated by a secret boiler. The whole operation was a scam.

Nevertheless, Springer expanded the compound to include a church, a 60-room hotel, a castle and an airstrip. But the feds were on to him and in 1974 he was arrested for misuse of public lands and for violations of food and drug laws.

Since 1976, California State University, Fullerton, has used the former health resort as a desert studies research center. Around the time of the conversion Buz and I checked the road for any possible snake hunting opportunities. But the pavement ended quickly and it was not common to find nocturnal desert snakes basking in the dirt. So we wrote off Zzyzx as an interesting but unsatisfactory snake hunting road, and we moved about thirty miles up Interstate

15 to Cima Road, which in our opinion has the best environment for snake hunting within at least thirty miles.

In 2008 Zzyzx was paved. That is, all of it was paved except for a small section which was left in its previous dirt condition in order to allow a herd of bighorn sheep to have access to a natural spring. The road is traveled quite a bit during the day, as the research center gets over 2,000 official guests each year. Some of these guests go back and forth on the road more than once a day. The area is also visited by many curious travelers, some of whom are simply fascinated with the name. Many of them park on the side of the road and pose for pictures next to the Zzyzx sign.

Because of all the traffic, it may not pay to cruise Zzyzx Road for snakes, but you never know. I'm leaning toward giving it another try.

The Strangest Road...

The strangest road in name and history in the Mojave Desert

CHAPTER 17
Cima – Near Perfect Mojave Desert Road

You can see the bright red light at the Shell gas station at the Cima Road off ramp for miles before you get there. If you are traveling on I-15 toward Las Vegas, it's the off ramp immediately after the last roadside rest. If you are heading southwest toward Baker, it's the first off ramp when you come down out of the mountain pass beyond State Line. You can't miss that red light, or the scrolling sign when you get closer inviting travelers to "See our waterfall urinal."

Well, I've seen it and used it, and it's decent as urinals go, but if all you need to do is take a leak, there's the roadside rest at the next off ramp, and there's a lot of sand on both sides of Cima Road. If that sign intrigues you, and you just have to see that waterfall, be prepared to part with some coin. They entice you in to enjoy a unique bathroom experience, but when you emerge you will find yourself surrounded by a vast assortment of homemade fudge, nuts, candies, unique drinks, souvenirs, sundries and supplies.

The first time Buz and I took that road, we went past the gas station and drove southeast toward the tiny town of Cima. It was dusk when we came to a sign with a blinking amber light that read, "Watch for Tortoise."

"See that, Buz," I said. "There is a sign telling us that reptiles are here. How often do you see anything like that?"

"We don't need a written sign," Buz replied. "Those Joshua Trees are all the signs we need. You know there's gonna be snakes in there."

Watch for Tortoises...

Watch for tortoises on Cima Road in the Mojave National Preserve

I discovered later that we were driving through one of the largest, Joshua Tree forests in the world.

Next we came to a sign that told us we were now in the Mojave National Preserve. We pulled over to have a look, and there was information about the area, and a permanent map.

Cima Road Section...

Cima Road section of the Mojave National Preserve

Darkness soon came and we continued our slow cruise (25 m.p.h.) along the two-lane desert road. There was no traffic at all. I don't believe it was more than a minute or two before we stopped for a Desert Banded Gecko, reflecting bright white in our headlights, just basking in our lane, almost begging to become the namesake of some of that jerky they sell just 26 miles down the interstate in Baker. This has always been one of my favorite lizard species. They are docile and delicate, almost transparent if you hold them up to a light. We snapped a few photos and walked this one off the road, releasing it back in the trees.

"Now we know there's two types of reptiles in here," I said. "All we need now are some snakes."

I was driving at this time. I thought music would be a good addition to the snake hunting experience, so I hit the search arrow on the radio. The first selection that came up was Elizabeth Cook singing *Sometimes it Takes Balls to be a Woman*. We both smiled as we listened.

Western Banded Gecko

Catch, Hold, Photograph and Release

A few minutes after releasing the gecko, a huge buck bounded out of the Joshua Trees and ran across the road only a few yards in front of the car. That woke me up, and I silently gave thanks that we were driving so slowly.

As we moved along I thought about that big buck, and contemplated the implications of *Sometimes it Takes Balls to be a Woman* when Buz yelled, "Stop!" I quickly stopped and he was out with a flashlight. Less than a minute later he was back with a beautiful night snake.

These fascinating snakes are considered rear-fanged and mildly venomous. Mind you, they are not dangerous to humans, and I have never seen one attempt to bite when handled. They have enlarged grooved teeth in the rear of their mouths on the upper jaw. They need to work a lizard's leg into the rear of their mouth, then hold on and chew in order to work the venom in. This has a paralyzing effect and allows the snake to swallow the prey.

Our specimen was especially nice, so we took it over to the gas station to photograph it under plenty of light. Later we brought it back and released it off the road.

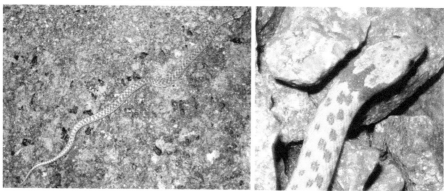

Cima Road Night Snake

Buz told me, "See, now we know there are three types of reptiles here. I'm telling you, Richard, when there are Joshua Trees like this, you're gonna find some snakes."

We found three other non-venomous snakes on that first outing. Rattlesnakes would be found on subsequent trips.

My favorite quality of this road for snake hunting is a general lack of traffic. There will be the occasional car or truck, but it is about as quiet a paved road as you can find. An interesting negative to this road is a severe drop in temperature that gets cooler the farther away you get from I-15. There is some kind of weird weather zone flowing through there that is much cooler than any part of the surrounding desert.

When the temperature is 100 in Baker it will be about 90 where I-15 meets Cima Road. By the time you traverse the entire length of the road (about 16 miles) it is generally in the low 70s or high 60s and that usually is not warm enough for snakes to come out and play. As you return toward I-15 it will warm up again. This is the only factor separating Cima from the truly great snake hunting roads.

As this road is part of the Mojave National Preserve, Buz and I would like to encourage herpers to seek and enjoy the animals, take pictures, help them off the road, but please don't take them home.

Snake Hunting East on Cima Road　　**Snake Hunting West on Cima Road**

Joshua Tree Forest

Where Joshua Trees grow, mainly in the Mojave Desert of California, Nevada, Arizona and Utah, the savvy herper is likely to find gopher snakes, night snakes, kingsnakes, glossy snakes, long-nosed snakes, Mojave Rattlesnakes, and others.

The name Joshua Tree reportedly came from a Mormon group crossing the desert. The trees reminded them of the Bible story of Joshua reaching up his hands in prayer.

The question is, are Joshua Trees really trees?

CHAPTER 18
Looking for Xantusias

Buz and I took our wives out to Mojave, California, to do some snake hunting. After the ninety-minute drive we stopped at White's Café for burgers and fries, some of their delicious homemade pies and steaming coffee. There was still too much daylight left to cruise for snakes, so Buz and I decided we'd go digging around in fallen Joshua Tree branches to see if we could round up a few night lizards, only we called them Xantusias, (we pronounced them *Zantuzies*). It was simply easier and more fun to refer to them by their genus, Xantusia.*

There are a handful of reptile genera that are entertaining to pronounce. Zantusia is one. Hypsiglena** (night snake) is another. A seldom used one that never fails to bring a smile to my face is Leptotyphlops (blind snake).

Desert Night Lizards are some of the smallest lizards in the United States, measuring only about 1½ inches from nose to vent, with tails about the same length. They are one of very few lizards that do not have eyelids. Contrary to popular belief, these lizards are neither rare nor strictly nocturnal. They are just secretive. They hide under, and especially inside of, fallen Joshua Tree branches, the more decayed the better.

At the time of this trip to Mojave, Buz was keeping a Hypsiglena, that is a Spotted Night Snake we had caught on the Ajo Road, between Ajo and Tucson. Unfortunately, this snake would only eat lizards.

Xantusias

Here is one of the night lizards we caught that evening.

Night lizards were perfect. Oak Creek Road had been a highly productive snake road for years, with perfect desert sand and lots of Joshua Trees. So we drove nearly its entire length until we passed the cutoff to the cement factory. Then we pulled over and parked. It was nearly dusk and it looked like it was going to be a beautiful sunset with an orange glow behind the mountains to the west.

Buz grabbed the coffee can that he always brought and a couple of screwdrivers, and we told Irisse and Sue that we would only be gone a couple of minutes. I explained that it would be dark soon and we wanted to be on the road looking for snakes. We were going to try to find a few Xantusias in the fallen Joshua Tree branches. Frankly, they were yakking and gossiping so intensely that I wasn't sure if they even heard me.

When I got to Buz, he was kneeling on a down slope where there was a fallen pile of decaying limbs. I watched him for a minute before carefully tearing into a thick, partially disintegrating branch of my own. I was less than thrilled about doing this because it's been my experience that you never know what is going to come out.

I don't mind the scorpions nearly as much as the beetles. With a scorpion you know what you've got. One time as I was tearing apart a length of branch a whole platoon of extremely shiny emerald green scarabs came pouring out like they would never stop. It freaked me out so badly that I almost puked, and I backed into some cholla and had to reach back and pull needles out of my ass for at least a couple of hours. Now I always check my surroundings for obstacles first, should it be necessary to make a hasty retreat.

This time we were clear of obstructions. Within a few minutes Buz had two Xantusias in the coffee can and I had one. It was getting dark, so we hiked back to the car. When the wives saw us coming they started laughing uncontrollably.

Buz just stared at them, but when the laughter began to subside I asked Irisse what was so funny.

In between giggles she managed to say, "While you were gone a cop pulled up and asked us if we were all right, and if we needed any help. Sue told him, 'It's all right, officer. Our husbands are just out there looking for Xantusias.'" Here they both started laughing again. The laughter was contagious and we joined in.

It was several minutes before Sue was able to say, "He just stared at us for a minute and then walked back to his car and took off."

Later that night we caught a long-nosed snake, a shovel-nosed snake and came across a dead-on-road Mojave Rattlesnake and Glossy Snake. It wasn't the best night of snake hunting, but not the worst either. We found some critters and the four of us had a lot of laughs.

Mojave Shovel-nosed Snake

We caught this on Oak Creek Road. We snapped this photo and walked the snake well into the desert to release it in the sand.

*For sake of clarity common names of animals are used in most cases in these stories, when in actual field activities scientific names are often preferred. It is more professional and more fun to understand, remember and use the scientific names.

**See "Suturing a Hypsiglena", Snake Hunting on the Devil's Highway

Ocotillo

It's always a good idea to learn your desert plants before you make a mistake, like trying to grab a desert iguana on one of these.

CHAPTER 19
Dreamy Draw

It was five minutes after midnight. Buz and I were hunting snakes in the Randsburg area some 35 miles northeast of Mojave, California. We had been cruising the roads for a few hours and had found a sidewinder, two long-nosed snakes, and one gopher snake. Nothing had been found for a half hour and I remembered something I wanted to tell Buz.

"Well, Buz, I'm going to turn the tables and tell you about a UFO incident that I heard about."

Buz was driving, and we were both concentrating on the dark road looking for a snake to reflect white in our headlights.

"Oh yeah?" he replied.

"Yeah, and this didn't come from a senile old rock shop miner dude with spittle running down his wild, tobacco-stained beard. My source is reliable. It's a guy who knows a guy whose bro is friends with someone who used to work at the Arizona-Sonora Desert Museum and is heavily involved with the UFO crowd."

"Well, yeah, sounds very reliable," said Buz sarcastically.

"So in October of 1947, just three months after the famous UFO crash in Roswell, New Mexico, an alien spacecraft crashed into the base of Squaw Peak, just outside Phoenix. A man named Silas Newton and another one known simply as "Dr. Gee," claimed to have pulled out two 4 ½ foot bodies and placed them in a freezer at a secret location. Later, officials from the U.S. Army picked them up. But they didn't cart off the remains of the spacecraft. Instead, they covered up the whole incident by building the Dreamy Draw Dam over the entire crash site.

"So I asked the guy whose bro is friends with a dude...well you know, the guy I told you about, why he thought it was that the government would leave valuable material from outer space laying around like that. He said that they already had recovered one just like it from the Roswell crash. That didn't sound exactly right, so I got interested in the story and did some research. I found out that they didn't build the dam until 1973. I asked the guy why they waited twenty-six years to complete the cover-up. He had two answers. First, he said that those in the know say this is a very deep and complicated conspiracy. And he said that many of his friends believe that the location where the dam was built has never been necessary for flood control purposes."

A big semi-trailer truck zoomed by in the opposite direction and both Buz and I shook our heads, hoping that we wouldn't find any furry or scaly road pizza ahead.

After a pause to let the story sink in I asked Buz what he thought. This is what he said: "I can believe the crash. I can believe the alien bodies. But I can't believe they would leave the crash out there unprotected. The military wouldn't do that unless they buried it, or disguised or camoed it. At Roswell they took everything."

We drove some more and he was quiet for a few minutes, absorbed in thought. Then he slowed down just a bit and turned to me.

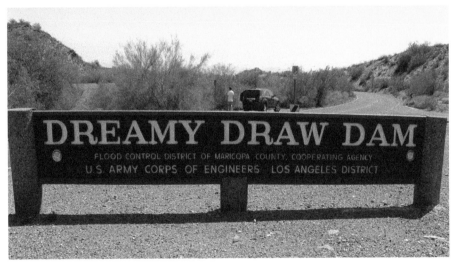

Dreamy Draw

Was this dam built to cover up a U.F.O. crash site, or was it built 26 years later to aid in flood control?

"Unless they really didn't feel they needed or wanted the remains. They could have diverted the water, you know, that they later built the dam for. If the crash was in a canyon, they could have blown up the side of the canyon and made the water flow into an "S" instead of a straight line. The material that came down from the blast could have covered up the crash debris. Then they would have time to get the dam built and a reason for doing it."

Then I remembered something. "What I forgot to mention," I said, "is that Dr. Gee and Silas Newton were later exposed as con-men."

"That's the first thing you said that makes perfect sense. That's how the guys in the dark sunglasses operate. They will gladly accept the alien bodies you are willing to remove from your freezer and hand over, but if you dare to tell anyone about it, you'll be exposed as a fraud and a con-man."

"Dreamy Draw sure has an interesting ring to the name," I said. "I'm getting tired. You up for calling it a night?"

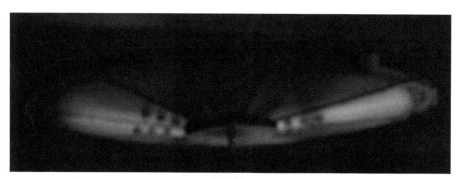

U.F.O.

This flying saucer can sometimes be seen in Sedona, Arizona.

CHAPTER 20
Don't Bug Me, Man

It began as an almost indiscernible brainwave. Then it became a thought, almost humorous and quickly dismissed. We had finished three nights of snake hunting and were beginning the long, 600 mile ride home. We were between Benson and Tucson, Arizona on Interstate Highway 10. I was driving, and Buz was taking a rare snooze. You can't get that guy to sleep while snake hunting, but he will sleep when the fun is over.

So I was driving west in the middle lane going about 70 miles an hour and being passed by cars and trucks alike. Then that almost humorous fraction-of-a-thought turned into a feeling, a horrible feeling of impending doom. Some hellish creature was ever-so-slowly crawling up my leg on the inside of my jeans. Whether it had 6 legs or 8 could not be discerned, but to me on that busy interstate, with a surprisingly large amount of midnight traffic, the leg count was matter of little importance.

I drifted into the slow lane and concerned myself with the thankfully light traffic moving along my driving leg, by then almost at my knee.

Buz was in a deep snooze. I had dozed a few times earlier when the snake hunting had slowed. Now I was wide awake, and struggling to concentrate on my driving, and on not freaking out from the little bug crawling like a moving tickle up my leg. Actually I did not know that this creature was little. I only knew that it was light.

I rolled the window in Buz's red and white International Scout partway down. The warm air felt cool on my feverish, sweaty face. A July monsoon had created a downpour earlier, and the sweet desert smell was still in the air. Horrible thoughts were going through my head at lightning speed. I imagined one of those tiny bark scorpions.

I knew enough about them to realize that if that thing cared to sting me on a testicle, if it didn't kill me, the pain and swelling would be so severe that I would yearn to be dead.

We were on the outskirts of Tucson now. I knew I would have to make a decision fast as the creature had crossed the intersection of my lower leg and my knee, and was now moseying along on the inside of my thigh. I wanted to scream, but it would have done no good. I made a quick decision and grabbed at the leg-traveler and had it corralled in a pinch-pocket of denim. Desert sweat was getting into my eyes, but now I didn't have a free hand to wipe it away. I took a deep breath and squeezed at the creature in the fold of cloth. I thought maybe I would puke if I heard a pop and felt an unnatural ooze on my inner thigh. But no. This monster had some kind of shell and would not pop in my tight finger grip.

Now what?

I pulled over into the emergency lane and rolled to a stop, still with that unearthly bug detained. In spite of everything I had the clear enough mind to look for the emergency flashers in Buz's Scout, but after a couple of seconds determined that they didn't exist or weren't important enough to keep looking for. I used the outside mirror to wait for a break in traffic. Eighteen-wheelers were whizzing by on their journey to get to their destination in Phoenix or San Diego or Los Angeles before dawn. I glanced over at Buz still asleep, leaning against the passenger door. Clearly he would not save me like he did when that loose Mojave Rattlesnake threatened my life in a similar manner several years before.*

When the moment was right I awkwardly opened the door, took another deep breath and made a dash for the open desert. I had no choice but to release my hold on the beast in order to undo my belt and get my pants off. Speed was exactly what was required in this operation. No fumbling around with shaking hands. Get the pants off, shake out the nasty beast, get the pants back on and get back in the car before becoming a spectacle.

Yes, that was exactly the ticket. But wait! What was this? I forgot that my cowboy boots had to come off first, and my feet were swollen. When I realized that I could not get the tight boots off standing up, and had no choice but to get down in the sand, my heart began pounding in my chest. I knew it had already been racing, but now it was going so fast that I was afraid it would interfere with my body's ability to function. If I were to have a heart attack right there and collapse, the bug might seek refuge in unspeakable places.

Next thing I knew I was on the ground using all my strength to push the boots off. It was during this process that I heard a voice break through the thick sound barrier of cars and trucks speeding by.

"Hey Rich, if you have to take a shit, there's a gas station at the next exit." Buz had awakened, rolled down the passenger window and was wiping his face with a bandana.

Not having time for a come-back, I popped off the left boot and struggled with the right until it was off. Then I stood up and hurriedly squirmed out of my pants. A wise-ass trucker laid on his air horn which further increased my heartbeat. I violently brushed off my entire leg, then the other one. Then I did a little dance and rubbed my entire body clear of any other real or imaginary creatures. Standing in my black underwear I shook out my pants. Even in the dim light I could see some kind of dark-shelled insect, almost the size of a dime flop onto the sand. Two or three cars honked as I quickly pulled my jeans up and fastened the belt. Before pulling my boots back on, I shook them out real good.

Trying to calm myself I turned to take one last look at the alien bug. Would you believe that that mutha effing thing gathered itself and flew away?

*See *Death by Mojave* in <u>Snake Hunting on the Devil's Highway</u>

Buz Quote

From Snake Hunting on the Devil's Highway

"A long time before the big government cover-up at Roswell, a spaceship crashed in this immediate vicinity. According to what I've heard, the craft had sailed directly over downtown Tucson and crashed into a windmill in a field right around here. The explosion scattered debris all around."

CHAPTER 21
Snap in the Box

Some time in 1969 I agreed to do a favor for a friend. I would pick up a snapping turtle from a location in the West San Fernando Valley, north of Los Angeles, and bring it, in its enclosure, to my friend's house in the Eastern part of the valley.

Sounds easy enough, and I usually don't mind doing favors for my friends, but in this case I hadn't asked enough questions, like how large was this snapping turtle? I had seen them as hatchlings in pet shops. That was what I pictured. Or maybe a slightly larger juvenile, like the size of my hand.

What kind of enclosure was it in? Would it easily fit in the back of my Volkswagen?

None of these questions occurred to me, and I thought this would just be an easy chore.

When I got to the address where I was supposed to meet a guy named Mike to pick up the snapper, there was a big cardboard box on the front porch. A note on the box said, "Sorry, had to make a run. Please take to Lee."

I looked around. There were no cars visible. It was a warm summer day and I was sweating. I pulled a flap back on the box, and couldn't believe my eyes. There inside, looking up at me was a large, adult snapping turtle, the legendary kind that some folks think could snap a broomstick in half with one mighty chomp.

I said, "Holy shit."

Could this be the specimen that my friend was expecting? And was this flimsy, cardboard box the enclosure that was mentioned to me? I would not call it an enclosure. It wasn't even taped shut. Only the flaps were turned in.

There were no cell phones in those days. I was not near any payphone, and I had no change anyway. I decided to ring the doorbell just to make sure no one was at home. The instant I pushed the button, at least two large dogs came charging the door inside the house, barking, growling, and, for all I knew, foaming.

I decided to take my chances with the snapper. I read the note again. "Sorry, had to make a run. Please take to Lee."

As I put the awkward, heavy box in the back seat of the Volkswagen, I wondered what kind of run this dude had taken. Groceries? Beer? Drugs? Well, it didn't matter. I didn't know the dude, and didn't think too much of him for placing what I estimated to be a fifteen to twenty pound potentially dangerous creature in a cardboard box with only the flaps tucked in.

It was so far so good when the snapper and I first started off on our voyage across the San Fernando Valley. I had no knowledge of the circumstances of this turtle's former existence. My expectation was that my friend would provide at least a better life for this turtle which really should have a huge plot of land, if not a pond or lake in its native environment. However, it occurred to me as I drove down Sherman Way, a major east-west thoroughfare that was lined with tall palm trees and had nothing but retail and office structures along its length, what if this turtle was raised from a hatchling? What if it had only known captivity? Would it adapt to the wild, if released where its species thrives?

I contemplated that question as I stopped for a red light. I heard some scratching in the box and turned around to look. Just as the light turned green, the turtle was effortlessly emerging from the box.

A teenager in a Camero laid on the horn and flipped me off as he passed on the right. I drove through the intersection and looked for a chance to pull over, but there was a line of cars in the right lane, and no way to pull over. Don't think for one second that any driver will slow down and let you move over if you put your blinker on anywhere around L.A. Nor was there anywhere to park in this area.

I drove along while the snapper busied itself getting a foothold on the top of the box so it could get all the way out. In the Volkswagen, its big beak was already just inches from me.

Five minutes later I was stopped at another light. The turtle was out of the box and trying to come to me. I don't know if intentions were to sit on my lap or snap off fingers. I only knew that this was the largest turtle I had ever dealt with up close and this personal.

When I was able to pull over, I had to get out of the car and open the front passenger door to grab the snapper by the upper shell. Fully expecting it to snap at my hands, I braced myself. But there was no attempt to bite. This snapper was as calm as a clam. I was able to get it back in the box, where it stayed nearly until I dropped it off. If I had had a place for this turtle, I would have wanted to keep it. It truly was a cool creature, once I learned it had no intention to eat me.

Lee was very happy to learn of its gentle disposition, and he provided a great area for it to thrive.

Petroglyph in the American Southwest

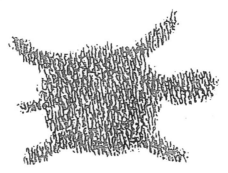

CHAPTER 22
The Extraterrestrial Highway

I have often wondered why, on snake hunting trips, Buz and I always seemed to get around to talking about E.T.s and cows.

Looking back, I thought of an early incident.

On one warm and breezy August late night we were cruising the roads south of Tucson looking for snakes. It had been over an hour since we found the last one. I needed to take a leak, and Buz wanted a cigarette and some thermos coffee.

Never mind that he had just related an urban legend about a crashed flying saucer in that very area! It didn't matter. I had an extra-long Magnum flashlight, like the police use, and before I unzipped to whizz in the Sonoran sandbox, I took my time and checked absolutely everywhere. Satisfying myself that there was not even a kangaroo rat or a dung beetle out there, I began taking a long-overdo leak. But in midstream, so to speak, an unearthly, blood-curdling MOO broke the stillness of the desert and interrupted my task in hand. Naturally I suspected Buz, but the sound came from out in the eerie desert, and not in the direction where Buz was sipping coffee and smoking a Pall Mall.

I have wondered (almost obsessed) about this incident for over 40 years. Then I stumbled upon three things that, along with a serious conversation with Buz, kind of, sort of, maybe-like-perhaps clarified this *cows and aliens* concept for me. At the very least it showed me that I was not the only person who marveled about the mystifying connection of aliens and cows.

The first was a slot machine. Living in Nevada is great if you like to take your recreation searching for snakes and from time to time going to the casino. I was walking by a row of slot machines one night and bam, my attention was instantly drawn to a machine called "Invaders from the Planet Moolah." I sat down and put in some money. I quickly learned that this is an alien-themed game where cows in flying saucers hover over a farm and abduct other cows. The player never learns why these alien cows are abducting other cows. All we know is the more cows that get abducted, the more money we can win. This is not anything like the reel slot machines we had when I was young, with bars and cherries.

The Invaders from the Planet Moolah slot was cool and entertaining, but it made me wonder, why did the makers conceptualize cows? I mean, why not pigs, goats, buffalo, armadillos or squirrels? I did some internet snooping and quickly found a YouTube video of a cow being beamed up into a flying saucer. Actually there are lots of these videos on YouTube. Did some of these short clips seem fake? Absolutely, but that didn't prevent that tiny seed from germinating in my dippy brain and sprouting question marks.

The topper was when Irisse and I took a day trip to the Extraterrestrial Highway, where I hoped to scout out herping* roads, and see sights like the so-called "Black Mailbox" and check out the Little A'Le'Inn Restaurant/Motel/Bar.

The Extraterrestrial Highway is only a pleasant 90 minute drive northwest of Las Vegas. Formerly designated only as SR 375, it was ceremoniously renamed in 1996 after many travelers reported strange activity and UFO sightings along this highway. For the purposes of my day trip, I was interested in the corridor going from Crystal Springs to the town of Rachel.

First thing I noticed is that the road looked promising for snakes. From Crystal Springs you drive up a gentle rise, and then go down into a valley. The road was straight and lightly traveled. Next I noticed cow signs every mile or so. The first one actually shows a cow being abducted into a space ship. That one really got my attention.

Speaking of signs, The Extraterrestrial Highway signs themselves are so full of graffiti they can barely be read.

Not far from the summit we found the location of the black mailbox, but it had been vandalized to such an extent that the rancher who owned it (we found by asking in the café) removed it. Now there is only junk in this well-known meeting place of sky-watchers.

We went into the Little A'Le'Inn and ate some very decent "Alien Burgers" for lunch in an alien-themed atmosphere. Outside, beside a small flying saucer being hoisted by an old jalopy tow truck, is a digital sign that warns drivers to beware of cows on the road.

When I returned home from the Extraterrestrial Highway, I couldn't wait to speak to Buz. I had been pondering the alien/cow issue for several weeks, and since Buz had an uncanny knowledge, and unending opinions about all things UFO, I was hoping that he would be the one to explain the relationship between cows and aliens, and put the subject to rest. That way, when we would next go snake hunting, he could find something else to freak me out about.

I visited Buz the very next day. He was recovering from a round of chemotherapy. This was not new to him. It seemed he had battled (and defeated) various cancers ever since returning from the Vietnam War in late 1967.

He was sitting back in his overstuffed chair with a faraway look in his blue eyes.

"How do you feel," I asked when I first walked in.

"Still with my fingers," he said.

I knew he would say that, but it was a way to bring him back from whatever medical world he was in at that moment. "I have some questions for you. I know how much you enjoy talking about UFOs and aliens."

Buz sat right up and his eyes cleared. A hint of a smile came brightened his face. "Sure," he said. "What do you want to know?"

I reminded him about the time I was taking a midnight leak while he was drinking coffee at the location of a rumored UFO crash.

"Do you remember that ungodly MOO?"

Buz started laughing. "You mean when you raced to the truck, frantically trying to zip up and yelling at me to get us out of there?"

I couldn't help chuckling, imagining my younger self doing exactly that. "Yes, Buz, that's the incident I'm referring to.

"So that was, for lack of a better word, the ghost of a cow, shrieking an alien-like *moo* near a reported crash site, not twenty minutes after you told me the gory details of that supposed crash.**

"So, Buz, why do you think we end up talking about UFOs so much on our snake hunting trips, and why is there so much talk of UFOs abducting cows these days?"

Buz smiled, and I braced myself for some sort of upcoming insult. "Well Richard, for over forty years I've watched you freak out over the tiniest insect or the mere mention of a bat or an E.T."

I knew he was right, but I also knew that I had come a long way over all these years. "I like bats now," I said with a chuckle. "So you had fun freaking me out. Is that why we almost always talked about UFOs? And what about cows?"

Buz lit a cigarette. "Great medicine for my C.O.P.D., huh?" He took a drag and exhaled a long stream of gray smoke. "Think about where we hunt, Richard. Much of it is around ranchland and much of it is open range. Think about all the cattle guards we go over."

"Right."

"Then, if you think about when we hunt in the desert, that's where the majority of the UFO activity is. Think about the Phoenix Lights, the crash in Roswell in 1947, the crash site near Tucson where you heard the space moo, and even the giant space figures I told you about, these were all in or very near desert locations."

I smiled and added, "Don't forget about the Dreamy Draw reservoir that they built to cover up that crash near Phoenix. Now care to share your opinions on cow abductions?"

Cow Abduction by Aliens Sign

"It's pretty simple really. I was gonna say it's not rocket science, but it kind of is. The grays***visit us from time to time to mine our resources. They take what they need to use or study. Many people have disappeared for a period of time, and when they return, they report that they were abducted, and usually admit to being probed. I ask you, Richard, would you admit to being probed by alien creatures from outer space if it didn't really happen? I know you wouldn't. And I doubt that you would say anything even if it did. So I think the likelihood is pretty high that happened to at least some of the many who have said so.

Some of the Open Range signs along the Extraterrestrial Highway have been updated to suggest cow abductions by aliens. Apparently some others of these signs (like a speed limit sign reported to have said "WARP") have mysteriously disappeared.

"Years ago there was a series of cattle mutilations. The most logical theory was sick cult members were doing this, but it was happening in several states, and even in other countries. Not long after that, entire cows were reported missing from farms. Sometimes, it was dozens of them at a time, usually from large open fields.

"If you want my opinion, and I know you do because you asked, it's gotta be for the beef or the milk, or both. They could have discovered the benefits of milk for their babies. And cows, either through their blood, organs, milk or meat could offer cures to alien diseases.

"Of course it could be that the grays just like burgers. I mean who doesn't?"

That made me smile. For the umpteenth time I asked him if he still wanted me to believe that he was an alien. "All these years I wondered if that was why we've been talking about all this stuff. Is that what you still want me to believe?"

"Of course! How else could someone like me know so much about aliens, reptiles and so many other mysteries of the universe?"

*Herping comes from herpetology, the science or study of reptiles and amphibians. Herping has come to mean the act of looking for, or hunting (not in the sense of duck hunting or deer hunting, neither of which I could ever do) but hunting to find reptiles and amphibians in the field or on the roads.

**See Snake Hunting on the Devil's Highway, pages 38-39

***Grays are another way of saying space aliens

Extraterrestrial Highway Sign

This sign is planted near Area 51

Extraterrestrial Highway Sign

Crystal Springs intersection of U.S. Highway 93 and State Route 375 (The Extraterrestrial Highway). This area is often used as a rest stop, although there are no facilities. From Las Vegas, this is the beginning of the 92-mile highway road that goes near the top secret Area 51, and through the tiny town of Rachel.

Welcome to Rachel Nevada sign.

Notice the flying saucer image near the top, and the two categories of population.

The Little A'Le'Inn Restaurant/Bar & Cabins

If you would like to have an Alien Burger or a couple beers surrounded by lots of alien-themed souvenirs, man, this is your place. Sit at the bar. Take a dollar bill out of your wallet and write your name on it. They will stick it above the bar with the thousands of others. The atmosphere is like a 50s diner collides with a real cool bar, and the whole thing gets zapped by an E.T.'s ray gun. The result is a palate that is as strange and interesting as the Extraterrestrial Highway is itself.

Tow Truck Hoisting Small Flying Saucer in the Parking Lot of the Little A'Le'Inn Restaurant/ Bar/Cabins

The Black Mailbox

First it was black. After it became the gathering place for UFO enthusiasts, and some of them were abusive, the owner, a local rancher replaced it with a sturdier mailbox that was painted white. It was still known as the black mailbox, and many a UFO had been cited in the area. Never mind that the facility known as Area 51 was right down the road from that spot. Eventually the white, black, sturdier mailbox was vandalized, and the owner, having grown sick and tired of the foolishness, removed it completely. Now there is just junk where the iconic mailbox once stood.

Buz and Richard Quote

From Snake Hunting on the Devil's Highway

Buz: Richard, it's just a cow!

Richard: Step on it, man. It's not a cow! It's an invisible, axe-wielding alien monster. It just moos to make you think it's a cow. I'm telling you, man. I looked all around out there and there was nothing!

CHAPTER 23
Return to the Sunizona Café

Sometimes one foolish act, carried out without considering the possible consequences, can be a life-changer. That's how it was with me forty-two years ago on my first trip to southeastern Arizona with Buz Lunsford. The owner of a desert café provoked and embarrassed me. I went out to the car and brought in a sack with a hot Mojave Rattlesnake, and after twice getting permission from that café owner, released the snake on the slick linoleum floor. That's when all hell broke loose.*

This act, so unlike anything I had ever done before, emboldened me afterward to write about the incident, and thus confess to the world that I was an idiot. Buz and I (mainly I) were lucky. No harm had been done to any person or animal that day, and in a large sense I had achieved revenge against a vile man who had taunted and belittled me in front of the many ranchers present in that café.

The article I wrote, *Diamondback Fever and Other Diversions*, was accepted by the *Arizona Republic* Newspaper, and published in the April 13, 1975 edition of their Sunday supplement called *Arizona*.

The results were unexpected. The Sunizona Café was visited by many curious readers. The proprietor eventually sold the café and each of the several subsequent owners knew the story of the two California snake hunters and the released rattlesnake. Each year when we returned for a few days of snake hunting, we were treated like celebrities. Always, someone in the café knew the story. People bought our lunch, shook our hands; I was even offered a teaching job by one woman owner who was on the local board of education. I thought that was crazy. How could they possibly consider a person like me to teach their children? Yet the gal was serious and even persistent, and for a short while, I actually considered the move.

Years later, a gal named Alice owned the café. She was a large woman with a heart the size of Texas, a real sweetheart. This was not so many years after singer/songwriter, Arlo Guthrie, came out with his iconic anti-war song, Alice's Restaurant Massacree. Buz was a Vietnam War vet. (see Chapter 9, *Snakes in the Grass*) I had protested that war in every way I could. Together, and ridiculously out of key, we sang Guthrie's words, "You can get anything you want at Alice's Restaurant."

What I wanted every time I entered that café was good food, and steaming fresh coffee in a clean and relaxing atmosphere. Alice made sure that my expectations (and those of her regular customers) were met.

A few years later, when the difficulties of being owner, cook, hostess, cashier, bookkeeper and janitor became too much for Alice to handle, she too sold the café. Buz and I were happy to find her cooking at the Willcox Truck Stop. We always enjoyed reminiscing with her, mainly because of her jovial composure, and we never had a substandard meal there.

The Willcox Truck Stop, located right off of Highway 10, Rex Allen Drive exit, and next door to the Best Western Plaza Inn, was where I was first introduced to salsa. It was homemade back in the 1970s, and I believe it still is to this day. I have had salsa all over the country, including places like Santa Fe, New Mexico, and El Paso, Texas, and I have never found any I enjoyed more than at this truck stop.

Circumstances kept us away from Willcox, and the Sunizona Café for over ten years. When we returned, we found the café out of business. We looked in the window and found chairs piled on top of tables in a dark room. The sign was down. It stayed that way for several years, possibly another small business casualty in a disastrous economy.

In 2016 I again visited Willcox, this time with my son, Jamie. Last time he was there, he was a pre-teen. Now he is grown and in his 40s. We drove to Sunizona and when we arrived at the intersection of Highways 191 and 181, I immediately noticed that the café was open once again. I pulled into the parking lot and stopped away from the building, just off the road. I got out of the car and looked around in all directions. Instantly my head was filled with racing thoughts and floating images.

I stood in the dirt parking lot, the same lot and likely some of the same dirt that was there four decades ago when I first went inside that café with Buz Lunsford. Turning around I faced the big highway. 40 years ago this was commonly known as *The Devil's Highway*, as it was enumerated 666. Pressure from certain groups caused the government to change every sign, every map, every single highway designation of 666, and replace them with the new number 191.

Buz and I had found many hundreds of reptiles, amphibians, mammals and other interesting creatures on that north-south road when it was Highway 666. For the record, most of the living critters we found were taken off the road and released. Many of our best adventures were recounted in <u>Snake Hunting on the Devil's Highway</u>. I stood and remembered when we observed hundreds, if not thousands of tarantulas migrating across this road not far from this very spot. I thought of the skunk feeding on a flipped over road kill snake, and how desperately we wanted to identify that snake, but the skunk wouldn't let us. I closed my eyes for a moment and I saw a temporary pond at midnight in front of the International Harvester tractor dealer south of here a few miles. In that pond were a dozen sets of four eyes. Pairs of toads from three different species were taking advantage of the water and mating while they could.

I snapped back to present day and watched a hay truck rumble north with a full load, likely coming up from Elfrida, where there are several huge alfalfa farms. Jamie was standing nearby smoking a cigarette and likely wondering what I was doing staring at the road. I pointed to the truck and said, "Hay there." He frowned, having heard those words all his life whenever a hay truck was in sight.

As I continued to watch the sparse traffic on the highway, so many thoughts and details danced in my head competing for attention. Not only had the number changed on this road, but it was different in other ways too. Down yonder not too far is a border patrol check point in the northbound lanes. It has been there for over 10 years. No matter what time of day or night, drivers must stop and speak to the officers. You can be searched for illegal aliens in your vehicle, or for drugs. Or for both. If you look or sound suspicious, the officers will not hesitate to bring out the dogs. That station did not exist back in the 70s and 80s, which was likely a good thing for us.

I turned and faced the café. The memories were so strong they almost knocked me back a step. I closed my eyes and, in speeded up motion, I heard a mixture of sounds: a hot and agitated Mojave Rattlesnake was buzzing loudly, the portly café proprietor was yelling, silverware was tinkling against ceramic plates, all conversations halting, wooden chairs sliding, and customers running for the walls. They did not head for the door, as the unpredictable rattlesnake was in their path. (I have thought from time to time that perhaps they wanted to stay and see the outcome of this most unusual event.)

My concentration was broken as a pick-up truck drove through the lot, kicking up a cloud of dust. Just as well. I decided to take pictures, not keep my son waiting and wondering as I relived everything that ever happened there.

Those days are as long gone as most of my hair.

That my life has been changed by events in this place is not debatable. I was young and foolish. I did something stupid, and it somehow worked out. It could easily have gone the other way.

I don't keep reptiles these days, except for the desert tortoises that I rescue and carefully re-home. Mind you, I never remove them from

Sandy's Restaurant From Highway 191

The old Sunizona Café is now Sandy's Restaurant. Highway 191 used to be Highway 666, the "Devil's Highway." The names have changed, but nothing can erase the history of what happened here over 40 years ago.*

the desert, and only accept the ones I'm able to keep when their owners no longer want them.

I still hunt for snakes in the dark of night, but now the snakes are all released off the road. I have developed a deeper appreciation for even the most common species. It's hard to describe my complete satisfaction from holding and beholding a wild snake, lizard, or any reptile or amphibian. It is even more fun to release them where they live. I like to think that I have learned a few lessons over the years by catching, and formerly by keeping snakes. The most important one to me, I think, is to help them live their lives naturally.

What used to be the Sunizona Café is now Sandy's. They have three flags hanging outside. One is a yellow *Don't Tread on Me*

Sandy's Restaurant, a Closer Look

Note the three flags. Don't tread on me features a rattlesnake. Hmmm. Wonder if they realize that they had one going crazy on their slick linoleum floor over 40 years ago. The black flag on the right symbolizes POW/MIA, (prisoners of war/missing in action) which became a major movement during the Vietnam War. This flag has great meaning for all of our veterans, including Buz. See "Snakes in the Grass" in this book. Of course the American Flag and both of the other two should be held in reverence by all Americans who love their country.

flag with a snake. In the middle is an American flag in red, white and blue. To the right is a black and white POW/MIA flag. After I snapped a couple pictures I stood up straight and thought about the meanings of each of those.

Jamie tapped on my arm. "Are we going in?"

I looked at the café, and turned to the road again. I glanced at my son standing next to me, now a grown man, much older than I was when I first stepped through those doors. I'm not sure what, but something made me smile. "Not right now, Jamie. Maybe later."

*See *The Sunizona Café* in <u>Snake Hunting on the Devil's Highway</u>

Arizona Police Officer Quote

(To Buz and Richard on Highway 666* near Pearce, Arizona as we were marking and releasing a rattlesnake)

"You know that 666 is the number of the beast, don't you? It comes from the Bible. Revelation 13:18 states, 'Let him who has understanding calculate the number of the beast, for it is the number of a man: His number is 666.'" And he said, "(People) believe that the Devil controls everything that happens on this road. And if they looked at you two right now, they'd probably use what you're doing to support their position."

Richard Lapidus

Hunting for snakes is like...

You don't know what you're going to find,
but you know it will be something good.

Paradise is located in the Chiricahua Mountains, Cochise County, Arizona, about 5 miles west of Portal.

Photo Courtesy of Troy Kelley, Johnson City, New York

Acknowledgements

My heartfelt thanks and deep appreciation goes out to Henry F. "Buz" Lunsford. There would be no interesting or amusing adventures without Buz, and therefore no book without his deep influence.

All of the photos in this book, unless otherwise stated, came from the archives of Herp-Ecology, an organization that Buz and I founded about 40 years ago.

I am deeply indebted to Peter Zuehlke, who once again sharpened up several old photos.

Thanks go out to my beautiful wife, Irisse. For many years she put up with up to 60 snakes in the house (she will swear it was 70), and hundreds of mice and rats in the garage. It's not so rare today, but back in the 1970s and 1980s it was asking a lot of a young wife with three toddlers in the house. 50 years later she still goes snake hunting with me.

Thanks go to my dear friend and prolific western writer, Phyllis de la Garza, who has never stopped encouraging me, but who will never let me slide if I don't tell a good story.

Thanks to Arizona's Official Historian, Marshall Trimble, for allowing me to borrow his published quotes about interesting places in Arizona.

Thanks to Troy Kelley for the use of the terrific photo of Paradise, Arizona.

Special thanks to Keith and Janice Davis and Goose Flats Publishing, Tombstone, Arizona. What an awesome experience it is working with you!

Finally, thanks go to my boys, Rory, Mickey and Jamie, who are my critics, public relations, web designers, proof readers, personal assistants and friends.

About The Author

Richard Lapidus has been a high school English teacher, a businessman and a writer. He is passionate about reptiles and the old west. His articles on these subjects have appeared in books, national magazines, major newspapers, history and college journals, and in books by other authors. He has been the master of ceremonies of a major western book event in Tucson, Tombstone and Willcox, Arizona, for nine years in a row. He is a former vice president of The Western Outlaw-Lawman Association (WOLA). He is a member of the Western Writers of America. Richard currently lives in Henderson, Nevada.

Other Books by Richard Lapidus

Snake Hunting on the Devil's Highway

Snakey Joe Post, Guardian of the Treasure

The Legend of Russian Bill

Fun With Snakes

CPSIA information can be obtained
at www.ICGtesting.com
Printed in the USA
FSHW01n0953110618
49092FS